Meiosis: Current Research. III.

Papers by:
John Melnyk, P. L. Pearson, Arthur Falek,
Alberto Solari, Petre Raicu, Helen Pogosianz,
Luciano Zamboni, Daniel Szollosi, Ford Calhoun,
Claes Ramel, R. C. Juberg, K. S. Lavappa,
Margaret Stevens, Aly Mohamed, J. et M. Moutschen-
Dahmen, Yasuo Hotta, S. K. Sen, Peter Beaconsfield,
Anil B. Mukherjee, Kathleen Church et al.

MSS Information Corporation
655 Madison Avenue, New York, N.Y. 10021

Library of Congress Cataloging in Publication Data
Main entry under title:

Meiosis: current research. III.

   The 2 earlier volumes of this 4 vol. collection issued under title: The meiotic process.
   Includes bibliographies.
   1. Meiosis. I. Melnyk, John.
QH605.M428   574.8'7623   72-6751
ISBN 0-8422-7036-1

Copyright © 1972
by MSS Information Corporation
All Rights Reserved

TABLE OF CONTENTS

Meiosis in Mammals	9
Failure of Transmission of the Extra Chromosome in Subjects with 47,XYY Karyotype .... Melnyk, Thompson, Rucci, Vanasek and Hayes	10
Fluorescent Staining of the Y Chromosome in Meiotic Stages of the Human Male .... Pearson and Bobrow	14
Meiotic Chromosomes of Man .... Falek and Chiarelli	19
Definitive Evidence for the Short Arm of the Y Chromosome Associating with the X Chromosome during Meiosis in the Human Male .... Pearson and Bobrow	23
The Spatial Relationship of the X and Y Chromosomes during Meiotic Prophase in Mouse Spermatocytes .... Solari	27
Distribution of Chromosomes in Metaphase Plates of *Mesocricetus newtoni* .... Raicu, Vladescu and Kirilova	47
The Evolution of the Ultrastructure of the Sex Chromosomes (Sex Vesicle) during Meiotic Prophase in Mouse Spermatocytes .... Solari	53
Changes in the Sex Chromosomes during Meiotic Prophase in Mouse Spermatocytes .... Solari	70
Meiosis in the Djungarian Hamster. I. General Pattern of Male Meiosis .... Pogosianz	78
The Behaviour of Chromosomal Axes during Diplotene in Mouse Spermatocytes .... Solari	90
Intercellular Bridges and Synchronization of Germ Cell Differentiation during Oogenesis in the Rabbit .... Zamboni and Gondos	104

Morphological Changes in Mouse Eggs due to Aging in the Fallopian Tube .... Szollosi	111
Genetic Effects on Meiosis	129
Genetic Analysis of Eight-spored Asci Produced by Gene *E* in *Neurospora tetrasperma* .... Calhoun and Howe	130
The Effect of the Curly Inversions on Meiosis in *Drosophila melanogaster*. I. Intrachromosomal Effects on Recombination .... Ramel	141
Etiology of Nondisjunction: Lack of Evidence for Genetic Control .... Juberg and Davis	149
Latent Meiotic Anomalies Related to an Ancestral Exposure to a Mutagenic Agent .... Lavappa and Yerganian	159
The Radiation and Drug Sensitivity of Meiotic Processes	165
Procedures for Induction of Spawning and Meiotic Maturation of Starfish Oocytes by Treatment with 1-methyladenine .... Stevens	166
Chromosomal Changes in Maize Induced by Hydrogen Fluoride Gas .... Mohamed	169
Meiotic Consequences of a Combined Treatment Fast Neutrons-FUDR in *Vicia faba* .. Moutschen-Dahmen and Moutschen-Dahmen	176
The Action of Phleomycin on Meiotic Cells ...... Hotta and Stern	181
Chromatin-Organisation during and after Synapsis in Cultured Microsporocytes of *Lilium* in Presence of Mitomycin C and Cycloheximide .... Sen	194
Oral Contraceptives and Cell Replication .... Beaconsfield and Ginsburg	199
Effect of 5-bromodeoxyuridine on the Male Meiosis in Chinese Hamsters (*Cricetus griseus*) .... Mukherjee	202
Meiosis in the Grasshopper: Chiasma Frequency after Elevated Temperature and X-rays .... Church and Wimber	204

CREDITS & ACKNOWLEDGEMENTS

Beaconsfield, Peter; and Jean Ginsburg, "Oral Contraceptives and Cell Replication," *The Lancet*, 1968, 1:592.

Calhoun, Ford; and H. Branch Howe, Jr., "Genetic Analysis of Eight-Spored Asci Produced by Gene E in *Neurospora tetrasperma,*" *Genetics*, 1968, 60:449-459.

Church, Kathleen; and Donald E. Wimber, "Meiosis in the Grasshopper: Chiasma Frequency after Elevated Temperature and X-rays," *Canadian Journal of Genetics and Cytology*, 1969, 11:209-216.

Falek, Arthur; and Brunetto Chiarelli, "Meiotic Chromosomes of Man," *American Journal of Physical Anthropology*, 1968, 28:351-354.

Hotta, Yasuo; and Herbert Stern, "The Action of Phleomycin on Meiotic Cells," *Cancer Research*, 1969, 29:1699-1706.

Juberg, R. C.; and Louise M. Davis, "Etiology of Nondisjunction: Lack of Evidence for Genetic Control," *Cytogenetics*, 1970, 9:284-293.

Lavappa, K. S.; and George Yerganian, "Latent Meiotic Anomalies Related to an Ancestral Exposure to a Mutagenic Agent," *Science*, 1971, 172:171-174.

Melnyk, John; Havelock Thompson; Alfred J. Rucci; Frank Vanasek; and Susan Hayes, "Failure of Transmission of the Extra Chromosome in Subjects with 47,XYY Karyotype," *The Lancet*, 1969, 2:797-798.

Mohamed, Aly H., "Chromosomal Changes in Maize Induced by Hydrogen Fluoride Gas," *Canadian Journal of Genetics and Cytology*, 1970, 12:614-620.

Moutschen-Dahmen, J.; and M. Moutschen-Dahmen, "Meiotic Consquences of a Combined Treatment Fast Neutrons-FUDR in *Vicia faba*," *Experientia*, 1967, 23:582-586.

Mukherjee, Anil B., "Effect of 5-bromodeoxyuridine on the Male Meiosis in Chinese Hamsters (*Cricetus griseus*)," *Mutation Research*, 1968, 6:173-174.

Pearson, P. L.; and M. Bobrow, "Definitive Evidence for the Short Arm of the Y Chromosome Associating with the X Chromosome during Meiosis in the Human Male," *Nature*, 1970, 226:959-961.

Pearson, P. L.; and M. Bobrow, "Fluorescent Staining of the Y Chromosome in Meiotic Stages of the Human Male," *Journal of Reproduction and Fertility*, 1970, 22:177-179.

Pogosianz, Helen E., "Meiosis in the Djungarian Hamster: I. General Pattern of Male Meiosis," *Chromosoma* (Berl.), 1970, 31:392-403.

Raicu, Petre; Barbu Vladescu; and Maria Kirilova, "Distribution of Chromosomes in Metaphase Plates of *Mesocricetus newtoni*," *Genetic Research* (Cambr.), 1970, 15:1-6.

Ramel, Claes, "The Effect of the Curly Inversions on Meiosis in *Drosophail melanogaster*: I. Intrachromosomal Effects on Recombination," *Hereditas*, 1968, 59:189-196.

Sen, S. K., "Chromatin-organisation during and after Synapsis in Cultured Microsporocytes of *Lilium* in Presence of Mitomycin C and Cycloheximide," *Experimental Cell Research*, 1969, 55:123-127.

Solari, Alberto J., "Changes in the Sex Chromosomes during Meiotic Prophase in Mouse Spermatocytes," *Genetics (Supplement)*, 1969, 61:113-120.

Solari, Alberto J., "The Behaviour of Chromosomal Axes during Diplotene in Mouse Spermatocytes," *Chromosoma* (Berl.), 1970, 31:217-230.

Solari, Alberto J., "The Evolution of the Ultrastructure of the Sex Chromosomes (Sex Vesicle) during Meiotic Prophase in Mouse Spermatocytes," *Journal of Ultrastructure Research*, 1969, 27:289-305.

Solari, Alberto J., "The Spatial Relationship of the X and Y Chromosomes during Meiotic Prophase in Mouse Spermatocytes," *Chromosoma* (Berl.), 1970, 29:217-236.

Stevens, Margaret, "Procedures for Induction of Spawning and Meiotic Maturation of Starfish Oocytes by Treatment with 1-Methyladenine," *Experimental Cell Research*, 1970, 59:482-484.

Szollosi, Daniel, "Morphological Changes in Mouse Eggs due to Aging in the Fallopian Tube," *The American Journal of Anatomy*, 1971, 130:209-226.

Zamboni, Luciano; and Bernard Gondos, "Intercellular Bridges and Synchronization of Germ Cell Differentiation during Oogenesis in the Rabbit," *The Journal of Cell Biology*, 1968, 36:276-282.

PREFACE

This multi-volume collection includes major experimental results in the field of meiosis published in English in the period 1968–1971.

Areas in which current research has developed rapidly in this field are meiotic DNA synthesis and the ultrastructure of the meiotic apparatus, particularly the synaptinemal complex. As elsewhere in contemporary biology, an intensive effort is being made to determine the basis of specificity, here the specificity of homologous chromosome recognition. The answer to this lies somewhere in the molecular structure of the synaptinemal complex, as papers in these volumes indicate. Meiotic DNA synthesis and repair have been convincingly associated with exchange of genetic material through cross-over. A great many variants of the meiotic process are now known and variants such as pairing, recombination and chromosome movement are described in these volumes.

Other papers presented deal with the study of meiosis as it occurs in man and other mammals, the radiation and drug sensitivity of meiotic events, the biochemical and physiological aspects of meiosis, and the effects on meiosis of particular genes, mainly in *Drosophila*.

Meiosis In Mammals

JOHN MELNYK.
HAVELOCK THOMPSON.
ALFRED J. RUCCI
FRANK VANASEK.
SUSAN HAYES.

FAILURE OF TRANSMISSION OF THE EXTRA CHROMOSOME IN SUBJECTS WITH 47,XYY KARYOTYPE

SIR,—The presence of two Y chromosomes in some men,[1,2] and tentative evidence from meiotic studies that transmission of the extra Y chromosome is prevented,[3,4] presents a most interesting problem in human sex-chromosome mechanisms. At least eighteen children have been reported[2] whose fathers had a 47,XYY chromosomal constitution, but none of these had sex-chromosome aneuploidy. In one family[3] the XYY father had seven children —six males and one female—all of whom were normal. Theoretically, following secondary non-disjunction, the progeny should consist of one XX, one XY, one XXY, and one XYY, if the extra chromosome segregates randomly. No reports of such families have appeared as yet.

Secondary non-disjunction of an autosome has been documented (in several reports) for only one aneuploid situation, that of G trisomy.[5] In females with Down's syndrome the observed transmission appears to confirm the expected 1/1 ratio in terms of normal and Down's syndrome progeny. In males with Down's syndrome, data are lacking for progeny studies; there is some evidence that individuals with the syndrome may be subfertile.[6]

Hypogonadism has been reported in some XYY individuals,[2] but in most instances the external genitalia are normal. The scarcity of family studies of XYY men therefore may be related to infertility, or to social instability which precludes marriage.

In order to evaluate the potential transmission of the extra Y chromosome in this unusual condition, meiotic

---

1. Jacobs, P. A., Brunton, M., Melville, M. M., Brittain, R. P. *Nature, Lond.* 1965, **208**, 135.
2. Brown, W. M. C., *J. med. Genet.* 1968, **5**, 341.
3. Thompson, H., Melnyk, J., Hecht, F. *Lancet*, 1967, ii, 831.
4. Parker, C. E., Melnyk, J., Fish, C. H. *Am. J. Med.* (in the press).
5. Lejeune, J. *in* Progress in Medical Genetics (edited by A. G. Steinberg); vol. III, p. 144. New York, 1964.
6. Reisman, L. E., Matheny, A. P. Genetics and Counseling in Medical Practice; p. 79. St. Louis, 1969.

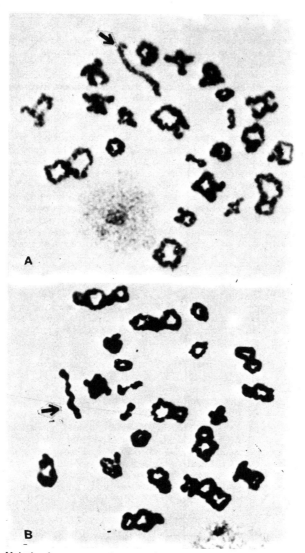

Meiotic chromosomes in two primary spermatocytes showing typical pairing of an X and a single Y chromosome:

(A) From a 46,XY patient. (B) From a 47,XYY patient.

studies were conducted on cells from six volunteers with a 47,XYY chromosome constitution. The patients, who were in an institution because of deviant sexual behaviour, were identified in a survey of two hundred tall men at Atascadero State Hospital in California.[7] The results of mitotic analyses of peripheral lymphocytes, skin, and testicular fibroblasts indicated that mosaicism was unlikely in these tissues. Chromosome counts were made on at least 25 cells from each tissue, and karyotypes confirmed a 47,XYY constitution.

Meiotic preparations from testicular biopsies were performed following modifications of the combined methods of Ohno[8] and Evans et al.[9] A total of 145 primary spermatocytes at diplotene and diakinesis were examined for XY pairing and for evidence of a second Y chromosome. On the basis of our experience with six meiotic studies from men known to have only one Y chromosome, it was possible to establish a subjective base line for the appearance of human XY bivalents. Against this background it was found that all of the sex bivalents seen in the XYY subjects had normal configurations which could be interpreted only as terminal associations between a single X and a single Y chromosome (see accompanying figure).

Cattanach and Pollard[10] reported the results of a meiotic and mitotic analysis of an XYY mouse, the first example of this condition in a mammal other than man. The presence of the extra Y chromosome as a univalent, or as a member of a YY bivalent, was demonstrated convincingly in this species. Evans et al.[11] described an XO/XYY mosaic mouse in which the extra Y chromosome appeared as a univalent in most of the XYY spermatocytes. The absence of the extra Y chromosome during meiosis in XYY humans therefore suggests a unique mechanism which eliminates the second Y chromosome prior to spermatocyte formation.

Ohno et al.[12] described the selective elimination of a sex chromosome during meiosis as a natural phenomenon in the sex mechanism of *Microtus oregoni*, the creeping vole. In this species the male, which has an 18,XY chromosome complement, selectively eliminates an X prior to spermatocyte formation, and produces some spermatazoa with a single Y and some without a sex chromosome.

---

7. Melnyk, J., Vanasek, F., Thomson, H., Rucci, A. J., Derencsenyi, A. *Nature, Lond.* (in the press).
8. Ohno, S. *in* Human Chromosome Methodology (edited by J. J. Yunis). New York, 1965.
9. Evans, E. P., Breckton, G., Ford, C. E. *Cytogenetics*, 1964, 3, 289.
10. Cattanach, B. M., Pollard, C. E. *ibid.* 1969, 8, 80.
11. Evans, E. P., Ford, C. E., Searle, A. G. *ibid.* p. 87.
12. Ohno, S., Jainchill, J., Stenius, C. *ibid.* 1963, 2, 232.

The mechanism which appears to eliminate the extra chromosome in man is not clear. Studies are now in progress and will be reported in due course.

FLUORESCENT STAINING OF THE Y CHROMOSOME IN MEIOTIC STAGES OF THE HUMAN MALE

P. L. PEARSON AND M. BOBROW

It has been previously reported that the distal half of the long arm of the human Y chromosome shows a differential affinity for fluorescent acridine derivatives (Zech, 1969; Pearson, Bobrow & Vosa, 1970). By the use of this staining reaction, the Y chromosome can be detected in interphase nuclei of lymphocytes, cultured skin fibroblasts and buccal mucosal cells. This technique has also been used to demonstrate that the X chromosome pairs with the short arm of the Y in first meiotic prophase (Pearson & Bobrow, unpublished observation). This paper describes the appearance of the Y chromosome in the different stages of spermatogenesis.

Cytological preparations for the examination of meiotic cells were prepared from testicular tissue of three normal adult males by the method of Evans, Brecon & Ford (1964). Some material was prepared without hypotonic treatment, particularly for examination of pachytene. Slides were stained for 3 min in a 0·5% aqueous solution of quinacrine dihydrochloride (Atebrin, G. Gurr), rinsed briefly in running tap water and then in distilled water, allowed to dry and then mounted in buffer pH 5·5. The specimens were examined on a Leitz Ortholux microscope with an Opak vertical illuminator, an HBO 200 light source with a 5 mm BG 12 excitor filter and a 510 nm barrier filter.

The Y chromosome can be positively identified in the great majority of lymphocyte mitotic metaphases, although about 10 to 20% of cells fail to show the typical staining reaction. The other chromosomal regions which commonly fluoresce are a small area next to the centromere of No. 3, and the satellites of one pair of D group chromosomes. In spermatogonial metaphases, however, the Y chromosome showed positive fluorescence in fewer than 5% of cells. The majority of nuclei showed no differential fluorescence (cf. Pl. 1, Figs. 1 and 2). In those cells which did exhibit a fluorescent Y chromosome, the intensity of fluorescence was less than that usually seen in other material. Interphase nuclei of spermatogonia did not show the fluorescent body seen in other interphase nuclei.

In all other stages of spermatogenesis, the Y chromosome fluoresced brightly. Plate 1, Figs. 3 to 6 show a succession of stages through first meiotic prophase up to and including pachytene. The fluorescent body is clearly visible in the great majority of cells at all stages. It can be seen to be associated with the 'sex vesicle', which is believed to contain the X and Y chromosomes. Smaller fluorescent areas, presumably representing the paracentric region of No. 3, can be seen in some cells (Pl. 1, Figs. 5 and 6).

At diakinesis, the sex bivalent regularly shows a terminal fluorescent area (Pl. 1, Fig. 7). Small fluorescent areas centrally situated on a large bivalent, presumably the No. 3 pair, are also shown in this figure. Where the sex chromosomes are not associated, the fluorescent area of the Y chromosome can be seen to extend for about half of its length, much as in mitotic preparations (Pl. 2, Fig. 8). The fact that the fluorescent area is always at the extreme end of the sex bivalent has been adduced as evidence that it is the short arm of the Y chromosome which pairs with the X chromosome (Pearson & Bobrow, unpublished observation). Plate 2, Fig. 10 shows an abnormal diakinesis. Despite the poor quality of the cell, a single structure with two large fluorescent areas can be seen. This is tentatively interpreted as a Y bivalent, the two Y chromosomes being present as a result of non-disjunction at a previous mitotic division.

The fluorescent area can be seen in some second meiotic metaphases (Pl. 2, Fig. 9), spermatids (Pl. 2, Figs. 11 to 13) and mature spermatozoa (Pl. 2, Fig. 14). The size of the fluorescent body in these highly condensed nuclei is comparable to that seen in somatic interphase nuclei. Spermatids may sometimes be seen lying in pairs, each pair presumably representing the products of a second meiotic division. In such cases, the members of the pair are almost invariably concordant for the presence or absence of the fluorescent body (Pl. 2, Fig. 11). In mature spermatozoa (Pl. 2, Fig. 14), the fluorescent body tends to be located at the acrosomal end of the nucleus, although many exceptions to this have been seen.

Theoretically, one would expect 50% of spermatozoa to contain a Y chromosome, although for technical reasons the actual proportion of cells showing a fluorescent body is unlikely to reach this level. Of 1420 spermatids scored from areas showing technically good staining, 733 showed no fluorescent body and

---

EXPLANATION OF PLATES

PLATE 1

Fig. 1. Spermatogonial metaphase; no fluorescent chromatin.

Fig. 2. Spermatogonial metaphase showing fluorescent Y chromosome.

Figs. 3 to 6. First meiotic prophase showing fluorescent area of Y chromosome associated with sex vesicle (large arrow) and fluorescent area of No. 3 chromosome (small arrow).

Fig. 7. Diakinesis. X—Y bivalent showing terminal fluorescence (large arrow) and chromosome No. 3 showing small fluorescent areas (small arrows).

PLATE 2

Fig. 8. Diakinesis. Separate X and Y chromosomes. Y chromosome showing differential fluorescence.

Fig. 9. Second meiotic metaphase, showing fluorescent Y chromosome.

Fig. 10. Abnormal diakinesis showing presumptive Y bivalent.

Fig. 11. Pairs of spermatids with and without fluorescent 'Y bodies'.

Fig. 12. Polyploid spermatid with two 'Y bodies'.

Fig. 13. Normal sized spermatid with two 'Y bodies'.

Fig. 14. Mature spermatozoa.

PLATE 1

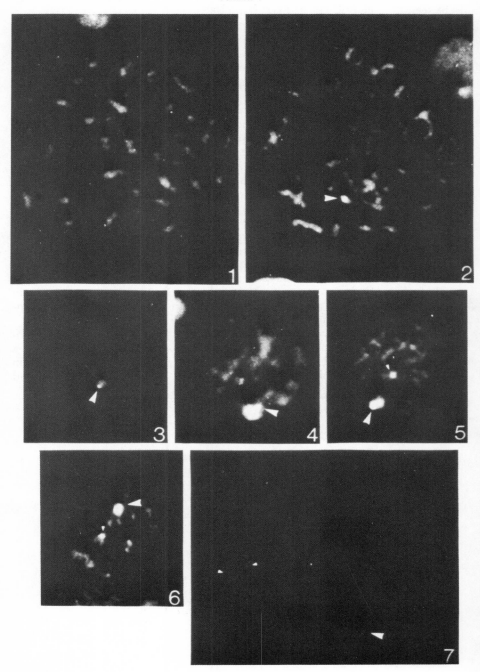

PLATE 2

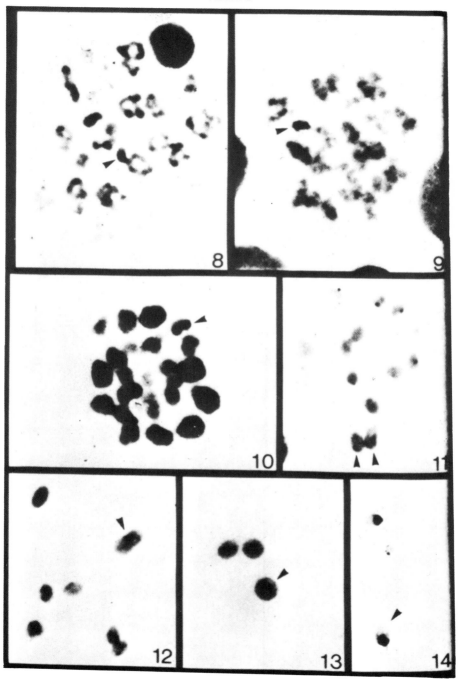

666 contained a single fluorescent body. The remaining twenty-one cells (1·4%) had two fluorescent bodies each. An example of such a spermatid is shown in Pl. 2, Fig. 13. Giant forms, presumably polyploid, have also been seen with two fluorescent bodies (Pl. 2, Fig. 12) but the twenty-one cells referred to above did not include any abnormally large forms. This proportion of gametes with two Y chromosomes cannot be immediately reconciled with what is known of the birth incidence of XYY males, but much more evidence is needed before the cause of this discrepancy can be discussed.

REFERENCES

Evans, E. P., Brecon, G. & Ford, C. E. (1964) An air drying method for meiotic preparations from mammalian testes. *Cytogenetics*, **3**, 289.
Pearson, P. L., Bobrow, M. & Vosa, C. G. (1970) Technique for identifying Y chromosomes in human interphase nuclei. *Nature, Lond.* (in press).
Zech, L. (1969) Investigation of metaphase chromosomes with DNA-binding fluorochromes. *Expl Cell Res.* **58**, 463.

Meiotic Chromosomes of Man

ARTHUR FALEK AND BRUNETTO CHIARELLI [1]

The meiotic chromosomes of man, in particular at diakinesis, have received increasing attention during the past few years (Book and Kjessler, '64; Finch et al., '66; Hulten et al., '66; Kjessler, '66; Lindsten et al., '65; Sasaki and Makino, '65). However, only slow progress has occurred in the detailed investigation of meiotic chromosomes of the human male since the actual haploid number of 23 was demonstrated by Ford and Hamerton ('56).

Our study is based on a small testicular biopsy, approximately 2 mm^3 in size, obtained from a 42 year old normal male treated for a hydrocoele. We are grateful to Dr. G. T. Cowart for providing this specimen which made our research possible. The technique employed to prepare the material for analysis is a modification of the standard aceto-orcein squash preparation described elsewhere (Chiarelli and Egozcue, '67; Chiarelli et al., '67). With this procedure we were able to obtain diakinetic chromosome spreads with almost all of the bivalents separated from one another.

In diakinesis the X and Y chromosomes appear linearly attached to one another, while the homologous autosomes are joined together to form bivalents. Of the 11 diakinetic metaphases analyzed, eight contained the expected 22 autosome bivalents and the set of sex chromosomes (fig. 1a). The other three spreads, probably as a result of our squash technique, were incomplete and showed 18, 19 and 21 bivalents.

To examine the morphology of the bivalents as well as determine the chiasma number for each chromosome, the chromosomes of the eight complete spreads were cut out and arranged in order according to decreasing size. One such arrangement is presented in figure 1b. As is evident, this is only a first step in the development of a procedure to compare specific meiotic chromosomes with those observed in somatic metaphase. In the establishment of this arrangement, in addition to size we were able to take into account some particular morphological details which appeared to characterize a specific chromosome. It can be reported that because of recognizable similarities and differences it was possible to arrange the chromosomes in the eight spreads obtained from our subject into what appear to be consistent patterns. The resemblance of particular chromosomes from spread to spread was due in large measure to the number and position of the chiasma along the chromosome filament.

The mean number of recognizable chiasma for each bivalent in the diakinetic spread ranges from 3.6 to 1. The first three chromosomes have a mean of greater than three chiasmata, chromosomes four through 16 have a mean between 2.1 and 2.9 chiasmata, and chromosomes 17–20 have two or less chiasmata. Chromosomes 21 and 22 can be readily separated from the other bivalents because of size and from one another by chiasma formation. A detailed examination of the number of chiasmata in each bivalent of the eight complete spreads revealed that the num-

---

[1] Dr. Chiarelli is extremely grateful to the Fullbright-Hays Commission which permitted him to stay at Dr. Falek's laboratory where this research work was carried out.

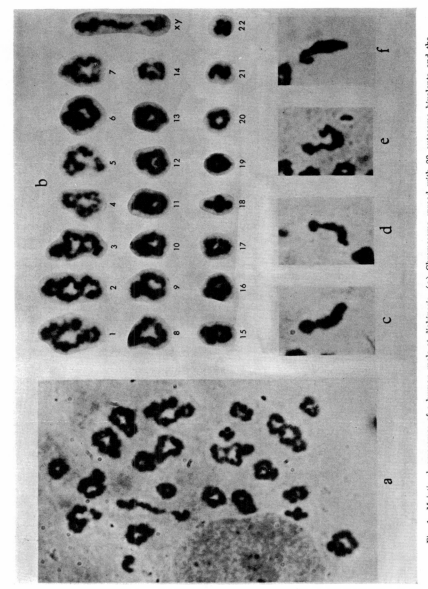

Fig. 1 Meiotic chromosomes of a human male at diakinesis: (a) Chromosome spread with 22 autosome bivalents and the separated X and Y chromosomes; (b) Karyotype of the spread arranged according to decreasing size and such particular morphological detail as number and position of chiasma; (c–f) Linear attachment of the X to Y chromosome in several spreads.

Chromosome number	Listed cells								Total
	A	B	C	D	E	F	G	H	
1	4	3	4	3	3	4	4	4	29
2	4	4	4	3	4	4	3	3	29
3	4	4	4	3	4	3	3	2	27
4	3	2	3	3	3	3	3	3	23
5	3	2	2	2	3	3	3	2	20
6	2	2	2	3	2	4	3	2	20
7	2	2	3	2	3	2	3	3	20
8	2	2	3	2	3	3	2	3	20
9	2	3	3	2	3	3	3	2	21
10	2	3	4	2	3	2	2	3	21
11	3	3	3	2	3	3	3	2	22
12	2	2	2	2	2	3	3	3	19
13	2	2	2	2	3	2	2	2	17
14	2	3	2	3	2	2	2	3	19
15	3	2	2	3	3	2	2	2	19
16	2	2	2	2	3	2	2	2	17
17	1	2	2	2	2	1	2	2	14
18	1	1	1	1	1	2	2	2	11
19	2	2	2	2	2	2	2	2	16
20	2	2	1	1	2	1	2	1	12
21	1	1	1	1	1	1	1	1	8
22	2	2	2	2	2	2	2	1	15
XY	—[1]	1	1	1	1	1	1	1	7
Total	51	52	55	49	58	55	55	51	

[1] X chromosome separated from Y.

Fig. 2 Chiasma record in eight cells with complete meiotic chromosome complements.

ber of chiasmata per cell is in good agreement with that reported by Ford and Hamerton ('56). The chiasmata counts for the spreads which contain all 23 bivalents are reported in figure 2. The total number in each cell ranges from 49 to 58 with figure 1b (Cell A, fig. 2) indicating 51 chiasmata. To evaluate further the importance of chiasma frequency to total chromosome number and speciation, it is planned to collect detailed information on a series of individuals in several mammalian species.

With regard to the X and Y chromosome the linear attachment and distinguishing characteristics between them are present in nearly all diakinetic figures examined (fig. 1c).

The Y chromosome appears as a short triangularly-shaped segment terminally attached at the end of the long arms to one of the extremities of the X chromosome (fig. 1d,e). In ten spreads the X chromosome appears terminally attached to the Y chromosome, while in one (fig. 1b) the Y chromosome is well separated from the X. The evidence of a true chiasma between the two sex chromosomes is still open to speculation (Koller, '37; Sachs,

'54). It may be that the chiasma between the X and Y chromosomes exists for only a short period at a very early stage of diakinesis. This may be the reason that it is not visible in many of the spreads. In the late stages of diakinesis the two chromatid composition of the Y is clearly seen (fig. 1d,e,f), but it is difficult if at all possible to detect the separation of the chromatids of the X chromosome. This consistent observation suggests a precocity in the time of the separation of the two arms of the Y chromosome in comparison with that of the X. The fact that the X chromosome often appears bent at a submedial point leads us to believe that it possibly represents the position of the centromere.

This investigation describes one possible way of classifying the chromosomes in diakinesis. Of significance is the fact that all the spreads utilized for this study appear to be in the same stage of diakinesis. This is based primarily on the observation of fairly consistent chiasmata counts and pattern shapes of similar chromosomes in different cells. Further support is lent by the fact that the smallest chromosomes with the least number of chiasmata showed no evidence of their terminalization. The introduction of such a procedure into the programs of cytogenetic laboratories will enable population studies of chiasma frequencies in the human male as well as investigation of the significance of such findings with regard to genetic variability and evolution.

In those individuals with anomalies due to chromosome rearrangement (translocation, deletion, inversion, etc.) the exact nature of the alteration could be recognized. Moreover, in mosaic persons the frequency of abnormal germinal cells could be evaluated. The importance of this finding for genetic counseling is evident.

LITERATURE CITED

Book, J. A., and B. Kjessler 1964 Meiosis in the human male. Cytogenetics, 3: 143–147.

Chiarelli, B., and J. Egozcue 1967 The meiotic chromosomes of two Macaca. Caryologia, in press.

Chiarelli, B., V. Mazzi and N. Beccari 1967 Tecniche microscopiche. Vilardi, Milano.

Finch, R. A., J. A. Book, W. H. Finley, S. C. Finley and C. C. Tucker 1966 Meiosis in trisomic Down's syndrome. Alabama J. Med. Sci., 3: 117–120.

Ford, C. E., and J. L. Hamerton 1956 The chromosomes of man. Nature, 178: 1020–1023.

Hulten, M., J. Lindsten, P. L. Ming and M. Fraccaro 1966 The XY bivalent in human male meiosis. Ann. Hum. Genet., 30: 119–123.

Kjessler, B. 1966 Karyotype, meiosis and spermatogenesis in a sample of men attending an infertility clinic. S. Karger, Basal.

Koller, C. P. 1937 The XY bivalent in human meiosis. Proc. Roy. Soc. Edinb., 57: 134–142.

Lindsten, J., M. Fraccaro, H. P. Klinger and P. Zetterquist 1965 Meiotic and mitotic studies of a familial reciprocal translocation between two autosomes of group 6–12. Cytogenetics, 4: 45–64.

Sachs, L. 1954 Sex linkage and the sex chromosomes of man. Ann. Eug., 18: 255–261.

Sasaki, M., and S. Makino 1965 The meiotic chromosomes of man. Chromosoma, 16: 637–651.

P. L. PEARSON
M. BOBROW

Definitive Evidence for the Short Arm of the Y Chromosome associating with the X Chromosome during Meiosis in the Human Male

MEIOTIC studies on the human male during the past thirty years, notably those of Koller[1] and Ford and Hamerton[2], have shown that the X and Y chromosomes associate at meiotic prophase. Whether the chromosomes form a true chiasma or not has never been demonstrated convincingly. Because the position of the centromeres of the chromosomes cannot be identified with certainty in meiotic preparations, there has been much speculation as to whether it is the long or the short arms of the Y chromosome which associate with the X. Thus, Sasaki and Makino[3] suggested that the $X-Y$ chromosome association involved the short arms of the Y, and Hultén et al.[4] suggested the long arms of the Y. The best cytological evidence has come from McIlree et al.[5] on a male with a dicentric Y chromosome involving a presumptive deletion of the short arms. At diakinesis the dicentric Y chromosome did not associate with the X, and Jacobs therefore suggested that in normal circumstances the Y chromosome might have associated with the X by its short arms[6]. From phenotype–karyotype correlations, Ferguson-Smith[7,8] deduced that part of the long arms of the Y

chromosome was homologous with part of the short arms of the X, and that this region was involved in the $X-Y$ association. From the appearance of the $X-Y$ association at diakinesis in testicular biopsies from a population of normal males, we have tentatively concluded that the short arms of the Y chromosome are involved in associating with the X chromosome.

Zech[9] has demonstrated that the distal ends of the long arms of the human Y chromosome at mitosis fluoresce

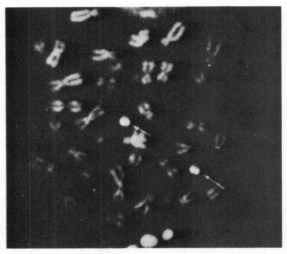

Fig. 1. Mitosis in a human male with XYY sex chromosomes showing a highly fluorescent distal region in the long arms of the Y chromosomes.

when stained with quinacrine mustard. The same effect can be achieved with quinacrine dihydrochloride[10] (Fig. 1) and the fluorescent Y chromatin can also be detected in interphase nuclei[11]. This can be used for determining the number of Y chromosomes present in the resting state nuclei of buccal mucosa cells. Moreover, the distal ends of the long arms of the Y chromosome fluoresce in meiotic cell preparations throughout all stages from spermatogonia through to mature spermatozoa[13]. We examined diakinetic cells prepared according to the technique of Evans et al.[12], stained them with quinacrine dihydrochloride[11] and viewed them with a Leitz microscope fitted with an HBO 200 light source and a 'Ploem' vertical illuminator. Fifty cells were selected from a normal male on the basis that they showed differential fluorescence and the $X-Y$ bivalent was examined under ultraviolet light and under phase contrast. In all cases where the X and Y chromo-

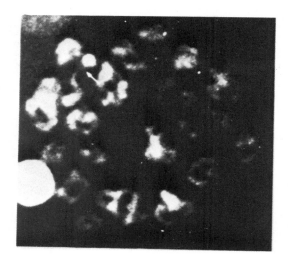

Fig. 2. **Diakinesis showing a highly fluorescent region** at the extreme end of the $X-Y$ bivalent in a normal male.

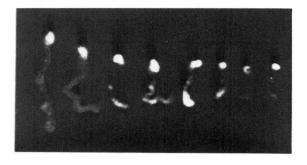

Fig. 3. Sex bivalents from different cells showing a fluorescent region at one end of each bivalent.

somes were associated, the more highly fluorescent region was situated right at one end of the sex bivalent (Figs. 2 and 3) and not in an interstitial position. Where the X and Y chromosomes were lying free, the Y chromosome characteristically had a highly fluorescent area and a less fluorescent area (Fig. 4), thus mirroring the staining distribution in the Y chromosome during mitosis. It

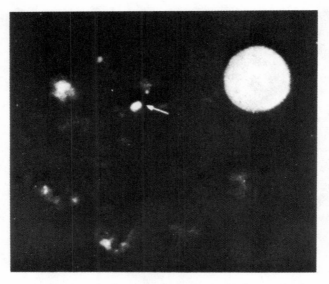

Fig. 4. Diakinesis showing a separate Y with a highly fluorescent and a low fluorescent region.

seems reasonable to suppose that the short arms, centromere and proximal parts of the long arm are situated in the less fluorescent part of the chromosome as they are in the mitotic Y chromosome.

We interpret these findings as definite evidence that it is the short arm of the Y chromosome which associates with the X during the first meiotic prophase.

[1] Koller, P. C., *Proc. Roy. Soc. Edinburgh*, **57**, 194 (1937).
[2] Ford, C. E., and Hamerton, J. L., *Nature*, **178**, 1020 (1956).
[3] Sasaki, M., and Makino, S., *Chromosoma*, **16**, 637 (1965).
[4] Hultén, M., *et al.*, *Ann. Hum. Genet.*, **30**, 119 (1966).
[5] McIlree, M. E., *et al.*, *Lancet*, ii, 69 (1966).
[6] Jacobs, P. A., *Brit. Med. Bull.*, **25**, 94 (1969).
[7] Ferguson-Smith, M. A., *J. Med. Genet.*, **2**, 142 (1965).
[8] Ferguson-Smith, M. A., *Lancet*, ii, 475 (1966).
[9] Zech, L., *Exp. Cell Res.*, **58**, 463 (1969).
[10] Vosa, C. G., *Chromosoma* (in the press).
[11] Pearson, P. L., Bobrow, M., and Vosa, C. G., *Nature*, **226**, 78 (1970).
[12] Evans, E. P., *et al.*, *Cytogenetics*, **3**, 289 (1964).
[13] Pearson, P. L., and Bobrow, M., *J. Reprod. Fertil.* (in the press).

The Spatial Relationship of the X and Y Chromosomes during Meiotic Prophase in Mouse Spermatocytes

ALBERTO J. SOLARI

Introduction

Moses (1958) described a tripartite structure that develops simultaneously with the synapsis of homologues during meiotic prophase. The general occurrence of this structure, the synaptonemal complex, during synaptic stages in many organisms has been reviewed by Moses (1964) and its possible involvement in crossing-over was supported by Meyer's (1964) observations. However, the presence of identical structures in single chromosomes (Sotelo and Wettstein, 1964; Moses, 1968) and the occurrence of polycomplexes (Roth, 1966; Moses, 1968) outside the synapsed chromosomes, shows that the synaptonemal complex cannot be simply equated to a specific structure of synaptic chromosomes.

The study of a single, identified chromosomal pair during its whole meiotic evolution may show some interesting data on the function of the synaptonemal complex. Thus, the length and duration of this structure could be measured, as well as its changes, and they could be correlated with the light microscopical observations on the same chromosomal pair and with genetical data.

The X-Y pair of the mouse has been previously studied in this laboratory (Solari, 1964; Solari and Tres, 1967a; Solari, 1969a, 1969b; Reader and Solari, 1969) and thus this pair has been chosen for a three-dimensional reconstruction at several stages of meiotic prophase. The aim of this paper is to present the results of the three-dimensional reconstruction of the X-Y pair at zygonema and pachynema and to discuss the significance of these results in relation to synapsis in the sex pair.

Material and Methods

Albino mice, random bred, were supplied from different sources. Two stocks from our laboratories were mainly used in these observations. Squashed spermatocytes did not show any alteration of the normal X-Y pair.

Electron Microscopy. Small pieces of testes were fixed in 2.5% glutaraldehyde in 0.1 M phosphate buffer, pH 6.9, for two hours, washed in buffer and postfixed in Caulfield's fixative for one hour. The pieces were embedded in maraglas. Sections were cut at different thicknesses, mainly three ones: thin (silver coloured), nominally 500—750 Å thick; intermediate (nominally 1,000 Å thick) and thick (nominally 0.47 μ thick). Serial sections were made with each of the three thicknesses, but the complete reconstructions of the sex pair here presented were done exclusively with thick (0.47 μ thick) sections. These thick sections present several advantages over the thinner ones: smaller number of sections needed, completeness of the cores in single sections and smaller number of measurements needed to determine their length. Thick sections, however, do not show sharp images as the thinner ones, and they need to be examined with higher acceleration voltages (100 or 80 kV in the Siemens Elmiskop). Thus, the complete reconstruction of 7 sex pairs was done with thick sections, and the reconstruction of special parts of the sex pair, like the common end, were done with thin sections to achieve a sufficient resolution at these special points. The observations made with the thinner sections agreed completely with those made with the thicker sections and they gave some additional details.

Serial sections were picked in single-hole grids (LKB Produkter) 1 × 2 mm in diameter according to Sjöstrand's (1967) method. The sections were stained with uranyl acetate (saturated solution in methanol) and then with Reynolds' lead citrate stain. Micrographs were taken at the same magnification step in each series, and then a diffraction grid (28,800 lines per inch) was photographed to calibrate the magnification.

Model Construction. Each micrograph from a complete series showing the whole sex pair was copied with china ink on a celluloid sheet (0.6 mm thick). Once a set of plastic copies is obtained (only the outlines of the sex pair, the cores and the nuclear and cytoplasmic membranes are copied generally) they are sequentially ordered like the negatives, and the points of coincidence between adjacent sheets are looked for. Generally the nuclear and cytoplasmic membranes, large vacuoles and nucleoli are the most useful markers. After coincidence is obtained with at least three markers, each pair of sheets is bound together with plastic tape and with a sandwiched number of slide pieces to serve as a spacer in the same scale as the micrographs. The process is repeated with the following sheets. Once a complete model is assembled, guiding holes are drilled in the edges of the whole model (three guiding holes are sufficient) to permit a rapid assembly of the model. Then the model is disassembled and the lengths fo the cores are measured in each sheet.

Measurement of the Cores. Lengths of each section of a core in each sheet are measured with a flexible copper wire (or with a map measurer). When the core is split into two filaments, one of them was chosen for the measurements. The measured lengths actually correspond to the projection of the core on the plane of the photographic plate. Each measured length is then a projection P_i where i is the ordinal number of the section. The nominal section thickness is constant in all the sections, $t = 0.47\,\mu$, and when it is multiplied by the scale factor gives T, thickness in scale. Then the following formula is used to evaluate the length of one core (L = total length of the core): $L = \sum_{i=1}^{i=n} l_i$ where l_i is the calculated length per one sheet and n is the number of sections; each l_i is calculated with the formula: $l_i = \sqrt{P_i^2 + T^2}$ when the core goes across the sheet; if both ends of the core leave the sheet by the same face, this special sheet (a marginal one for this core) is calculated with the formula: $l_i = 2\sqrt{\left(\frac{P_i}{2}\right)^2 + T^2}$. The error inherent in this method of length calculation is a systematic error that will be the same for measurements in different cells thus permitting comparisons of measurements during different stages.

Time Labelling of the Meiotic Stages. An adequate knowledge of the meiotic stage is needed for the analysis of the changing relationships of the XY pair. The time sequence was made as previously described (Solari, 1969a), that is, based on the stage of the spermatogenic cycle in each section as seen with low magnification and correlated with thick sections observed with the light microscope.

Results

1. Reconstruction of the X-Y Pair at Zygonema

Fig. 1 shows the characteristic stage 12 of mouse spermatogenesis with meiotic metaphases and anaphases in the upper spermatocyte layer. The lower spermatocyte layer shows cells in zygonema; the one selected for reconstruction is marked by the arrow. A series of thick (0.47 μ thick) sections of this cell showing the whole path of the cores of the sex pair is shown in Fig. 2a—g. Two cores are found in the sex pair (Fig. 2g). The long and the short cores are paired at one end, forming a rather long synaptonemal complex about 2,000 Å thick (Fig. 2b and c). The central element of this synaptonemal complex is found between the cores in the last half of this pairing region (Fig. 2b) but it may extend along most of this region. The synaptonemal complex thus formed is about 1.9 μ long and it is curved as an arc that lies approximately on a single plane. When both cores separate from each other, the short core has a short path that ends on the nuclear membrane. The long core goes across the central mass of chromatin of the sex pair (which is distinguished by its homogeneous degree of condensation) forming a large loop. The long core is thicker in this loop than where it forms the common end region; and in some parts of the loop it is clearly double, formed by two dense filaments about 300 Å each in thickness

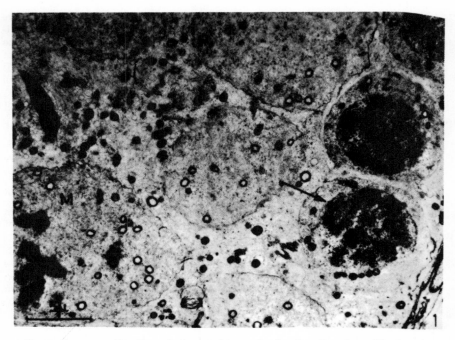

Fig. 1. Low-magnification electron micrograph showing the stage 12 of mouse spermatogenesis, with meiotic divisions (M) in the inner layer of spermatocytes; spermatocytes in zygonema (Z) are in the outer layer. The sex pair selected for reconstruction (Fig. 2) is marked by the arrow. × 3,400

(Fig. 2f). Then the long core curves towards the nuclear membrane and ends on it, surrounded by a mass of strongly condensed chromatin about 0.7 µ in diameter (Fig. 2a). Thus the general pattern of the two cores (Fig. 5) is already developed in zygonema. The two cores have a common end region forming a synaptonemal complex which is longest at this stage. Three regions can be distinguished in the chromatin of the X-Y pair at zygonema: the chromatin enclosed between the short core and the nuclear membrane, as well as the chromatin that surrounds the free end of the long core, are more packed than the main mass of chromatin (the one that corresponds to the large loop of the long core).

2. Reconstruction at Early Pachynema

In all the four cells reconstructed at early pachynema (stage 2—3 of the spermatogenesis) the pattern of the cores was similar to that of zygonema (see Fig. 3 in Solari, 1969b). However, the synaptonemal complex formed at the common end region is shorter than in zygonema

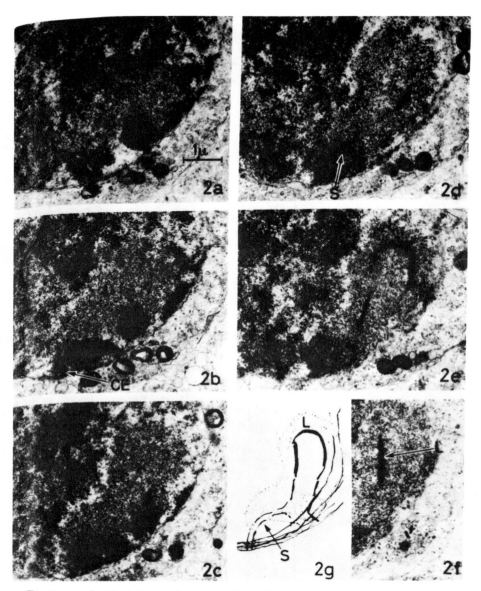

Fig. 2a—g. Serial electron micrographs from thick (0.47 μ) sections showing the whole path of the long (L) and the short (S) cores of the sex pair at zygonema. CE common end region. Fig. g is a photograph of the model. × 9,800

(see the Table, p. 225). The chromatin of the free end of the long core remains denser than the main mass of chromatin, but the chromatin related to the short core is no more distinguishable from the main mass.

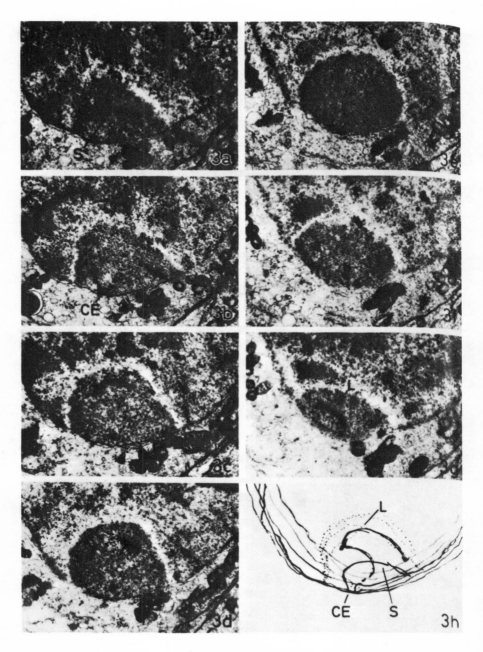

Fig. 3a—h. A series of electron micrographs from thick (0.47 μ) sections showing the whole path of the long (*L*) and the short (*S*) cores of the X-Y pair at the beginning of mid-pachynema. *CE* common end. The long core (*L*) is shown double at many places. Fig. h: photograph of the reconstructed model. × 8,400

3. Reconstruction at Middle Pachynema

A sex pair reconstructed at mid-pachynema (stage 6 of the spermatogenesis) is shown in Fig. 3a—h. The basic pattern of the cores is conserved but some significant changes have occurred. The long core is more convoluted and its two components are clearly separated in most of its length except at the common end (Fig. 3a—f). The distance between these two components equals the distance (about 1,000 Å) that separates the lateral elements of the autosomal synaptonemal complexes. Actually, a central element is visible at some points of the long core in the thick sections. This fact was confirmed with thin sections (Figs. 7 and 8) that showed the existence of an anomalous synaptonemal complex.

The common end is very shortened, but it is clearly distinguishable (Figs. 3b and 6a and b). Its structure is described below. There are no longer special parts of the chromatin mass distinguishable by a special condensation. Thus, during zygonema and early pachynema the regional condensations dissapear. During this stage a mass of dense, 200 Å thick granules begins to cover a part of the inner side of the X-Y pair, always in relation with the chromatin that surrounds the long core.

4. Reconstruction at Late Pachynema

During this stage the pattern of the cores remains the same, but some changes are reversed. Thus, the long core does no longer show the structure of a synaptonemal complex that appeared during mid-pachynema; both filaments composing the long core have now approached each other, and the cross-section of this core has often an oval profile suggesting a coiling between these filaments. The nucleolus is strongly developed around the chromatin that is centered by the long core (Fig. 4a and b). This relationship between the long core and the nucleolus is enhanced by the presence of invaginations of nucleolar material that come near, but do not touch the long core (Fig. 4b).

The common end region exists as a very changed part of the cores (Fig. 4a). It is very short (about 0.3 μ) and the ends of the short and the long cores have fused together in a dense rod (Fig. 4a) where four denser lines are seen (in the thin section shown in Fig. 4a only two lines appear). From this fused common end region both the short and the long core come out straightly and almost orthogonally to each other (Fig. 4a).

5. General Features of the Cores of the X-Y Pair

As seen in all of the reconstructions, there are two cores running inside the chromatin part of the sex pair ("sex vesicle"). These cores never enter the nucleolar part when it is developed during middle and late pachynema. The two cores have constant differences thus permitting

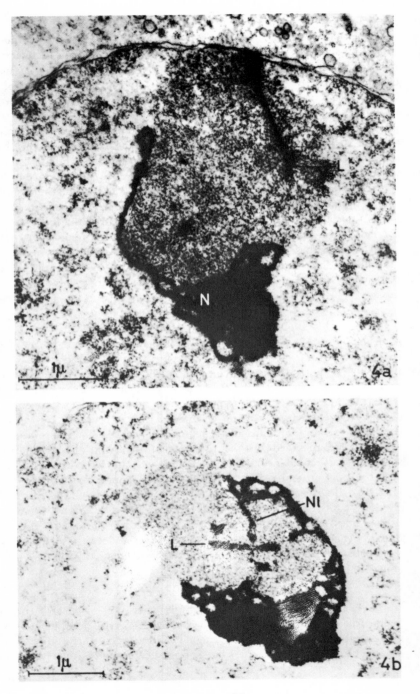

Fig. 4a and b

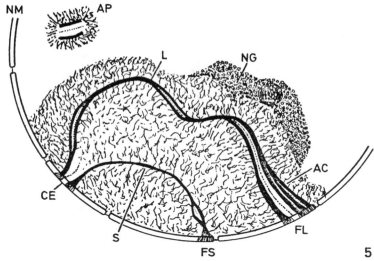

Fig. 5. Schematic drawing showing the X-Y pair during mid-pachynema. *L* and *S* long and short cores respectively. *CE* common end with a short synaptonemal complex. *FS* free end of the short core; *FL* free end of the long core; *AC* anomalous synaptonemal complex developed on the long core; *NG* nucleolar granules, *AP* autosomal pair; *NM* nuclear membrane

their identification. As they differ greatly in length, they are identified as the *long* core and the *short* core (Figs. 5, 2g and 3h). The length ratio of the short versus the long core, 1:2.5, is rather constant during prophase (Table). The absolute lengths of the cores do not show great

Table. *Measurements of the cores of the X-Y pair at several stages*

Model No.	Stage	Long core (μ)	Short core (μ)	Length ratio (long/short)	Common end length (μ)
1	mid-pachynema	8.3	3.7	2.2	0.2
2	zygonema	8.2	3.4	2.4	0.95
3	early pachynema	8.6	3,1	2,7	0,54
4	early pachynema	9.7	3.6	2.6	0.43
5	early pachynema	9.9	3.9	2.5	0.40
6	late pachynema	8.2	—	—	—
7	zygonema	9.9	3.2	3	1.9
Average		8.9	3.5	2.5	

Fig. 4a and b. Electron micrographs of thin (700 Å thick) sections of the X-Y pair at late pachynema. The nucleolus region (*N*) is completely developed and shows invaginations (*NI*) towards the long core (*L*). In Fig. a the common end region is a short, dense rod ending on the nuclear membrane, from which the short (*S*) core and the long (*L*) core come out straightly. × 22,500; × 20,000

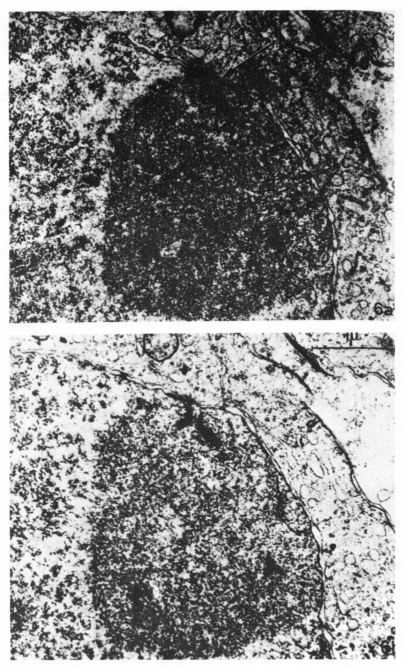

Fig. 6a and b. Electron micrographs of two serial sections showing the common end and its short synaptonemal complex (arrow) at mid-pachynema (section thickness, 1,000 Å). The long core (*LC*) is double from the common end. × 22,500

changes during the examined stages, although the measurements suggest that a little shortening of the long core can occur during these stages. Both cores begin and end on the nuclear membrane, like the autosomal synaptonemal complexes.

There are three ends of the cores: one common end of both cores and one free end from each of them.

6. The Common End

This region of the cores is formed by the approachment and the association of the long and the short cores a little before ending against the nuclear membrane (Figs. 6a, b, and 5). During zygonema, early and middle pachynema this common end region is a synaptonemal complex in which each lateral element is formed by each core. After the common end region both cores separate from each other almost in perpendicular directions. The length and structure of this common end varies during prophase but it is present from zygonema to diplonema (it has not been identified after diplonema). The length of the common end shortens progressively from a maximum of about $1.9\,\mu$ to a minimum at late pachynema of about 0.25—$0.35\,\mu$, that remains during diplonema. The common end shows the three elements of a synaptonemal complex up to mid-pachynema (Fig. 6a and b), the only characteristics being the duplication of one of its lateral elements as soon as it begins to separate from the other (the double core is the long core) and some variations in the path of the central element, which becomes curved or oblique and can deviate towards the end of one of the lateral elements. After mid-pachynema the space separating both lateral elements and the central element disappears (Fig. 4a).

7. The Free End of the Long Core and the Anomalous Synaptonemal Complex

The double structure of the long core becomes more evident during mid-pachynema when the two filaments composing this core become more separated (up to 1,000 Å) and a central element appears between them (Figs. 7 and 8b). This separation occurs at several places of the long core, possibly in most of its length except the common end region, but it is more evident at the free end (Fig. 7). At this point, and exclusively during mid-pachynema, a long stretch of the long core forms a very curious, asymmetrical structure: the anomalous synaptonemal complex (Fig. 7). While one of the lateral elements of this synaptonemal-complex-like structure is similar to the normal ones, the other is thicker, about 1,500 Å wide. This thicker lateral element is composed of at least three dense filaments about 300 Å wide surrounded by a fibrillar matrix. The

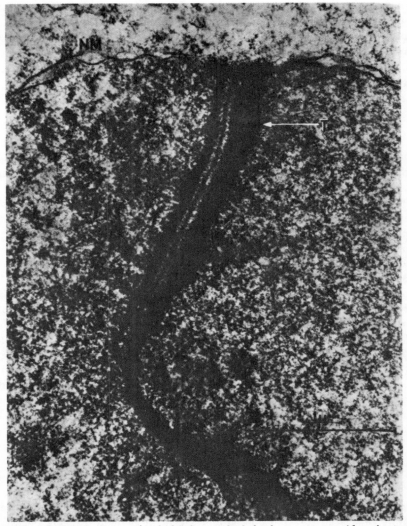

Fig. 7. Electron micrograph of the free end of the long core at mid-pachynema showing a long stretch of the anomalous, asymmetric synaptonemal complex with its thick (T) lateral element. × 40,000

central element is similar to that of an autosomal synaptonemal complex. This asymmetrical structure is more easily seen at the free end but it is present at least in the final third of the long core. The thicker element of this anomalous synaptonemal complex faces the nucleolar mass and

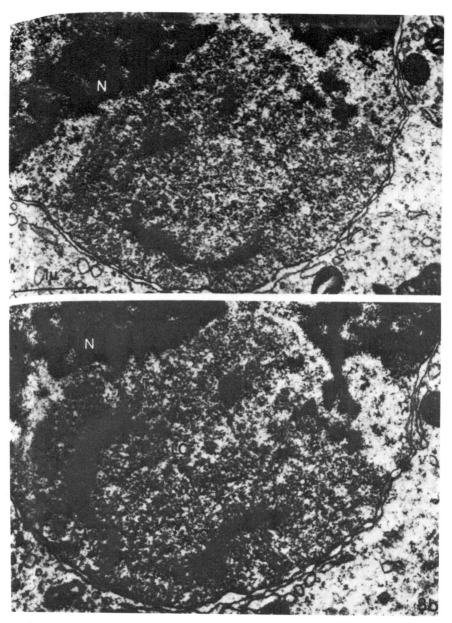

Fig. 8a and b. Two serial thin sections showing the anomalous synaptonemal complex of the long core (*LC*) at advanced mid-pachynema, and the relationship of the thicker lateral element (*T*) with the developing nucleolus (*N*). × 23,000

some nucleolar invaginations come near to it (Fig. 8a and b). The chromatin fibers that join both lateral elements are indistinguishable from each other.

While the short core does not form any synaptonemal complex by itself at any stage, it is split into two filaments that rejoin after a short distance near its free end during mid-pachynema, forming an "eye-like" zone (Fig. 3b).

Discussion

1. The Meaning of the Cores of the Sex Pair

As shown previously (Solari, 1969b) and in this paper, the long and the short cores are constant and characteristic elements of the sex pair of the mouse during meiotic prophase. In the human (Solari and Tres, 1969), and in the rat (in preparation) the same two cores (with different and specific length ratios) have been found. The length ratio of the cores remains rather constant during zygonema and pachynema, although the cores become more folded and suffer other changes. This constancy suggests that the cores represent — or that they are involved with — some fixed structure of the chromosomes.

During early zygonema the long core is axial to a rod-like mass of chromatin, that is connected by one end with a smaller mass of chromatin. The short core is axial to this smaller mass of chromatin (Solari, 1969b; in some cells this disposition of the cores is observed up to the beginning of pachynema).

At late pachynema the cores separate straight from each other as if pulled from their free ends. Thus, the characteristics and the behaviour of the cores are similar to those of a hypothetical chromosomal axis of each chromosome. Besides, the length ratio of the cores is nearly the same as the length ratio of the Y and X chromosomes observed with the light microscope (Ohno and Lyon, 1965). From these data we conclude that the cores are axial structures of each sex chromosome during zygonema, pachynema and diplonema.

The formation of the cores in the sex pair seems to be simultaneous to the appearance of cores (lateral elements) in the autosomes during early zygonema (Solari, 1969a). During leptonema well-defined cores are not present in the mouse, but they have been described in other species (Moens, 1968).

However, the evolution of the cores of the sex chromosomes differs from those of the autosomes. While the cores of the sex chromosomes become paired only at the common end region, synaptonemal complexes are formed all along the paired autosomes, as shown in *Gryllus* by Wettstein and Sotelo (1967) and in the mouse by Woollam, Ford and Millen

(1966). In this respect, the cores of the X-Y pair behave like the cores of the unpaired sections (univalents) of the chromosomes in the triploid *Lilium tigrinum* (Moens, 1969). The cores of the sex pair remain thick and dense during diplonema, when the autosomes offer a different picture.

In Searle's X-autosome translocation, the core of the translocated segment of the X chromosome combines itself with the core of one autosome outside the sex pair (Solari, 1969b), thus mirroring the behaviour of the axes of the chromosomes involved in that translocation.

Thus, the available evidence suggests that the cores are axial structures of the sex chromosomes, formed in the same way as the lateral elements of autosomal synaptonemal complexes (this suggestion was advanced earlier, Solari, 1964).

2. The Cores and the Structure of the Sex Bivalent

While the cores can be related to the linearity of the sex chromosomes, their position with respect to the two chromatids composing each homologue shows further complexities. Both the X and the Y chromosomes are composed of two chromatids at diplonema as seen with the optical microscope (Ohno, Kaplan and Kinosita, 1959). DNA synthesis at premeiotic interphase duplicates the DNA content of the nucleus (Rhoades, 1961), thus showing that the essential part of the chromosomes is already double at meiotic prophase. A small quantity of unreplicated DNA, as that described in *Lilium* by Stern and Hotta (1969) (0.3% of the genome) would probably not be sufficient to prevent the expression of chromosome doubleness.

Thus, the cores of the sex pair may be related in one of the two following ways with respect to chromatids: either the cores can be related to the axis of each of the chromatids, or they are related to the axis of the whole chromosome. The first alternative seems less probable, as the short core is not clearly double along its whole length. Furthermore, the doubleness of the long core, that is enhanced during mid-pachynema, is lessened during late pachynema and diplonema, contrary to what could be assumed to happen according to the light microscope image of the chromosomes. The most significant proof of the relationships between chromatids and cores would be given by their involvement in chiasmata formation. However, this question requires further genetical and structural data. On the whole, the second hypothesis seems more probable. The place that can be occupied by the core if it is co-linear to the whole chromosome but is not axial to each chromatid, can be the interface between chromatids. If the cores are placed at that interface, they do not need to be double, and they can remain in place after separation of the homologues during diplonema while the chromatids remain approached to each other. According to this hypothesis the cores do not necessarily take part in the formation of chiasmata.

A second question remains about the relationship of the cores with the basic structure of the chromosomes, that is, the chromatin fibrils. The core can be interpreted as a part of the fibrillar components of the chromosome, packed in a special array and density. The apparent continuation of the chromatin fibers with the substance of the cores seems to support this idea. However, Coleman and Moses (1964) and Moses (1968) have shown that the denser part of the lateral elements of synaptonemal complexes (which are similar to the cores of the sex pair) is deceptively poor in DNA to mirror a packed array of DNA-containing fibers. Furthermore, the cores of the XY pair diminish in staining affinity by extraction with cold perchloric acid, a treatment which does not affect DNA but extracts RNA and some basic proteins (Solari and Tres, 1967a). Evidence has been provided by Sheridan and Barrnett (1969) about the richness in basic protein in the lateral elements of synaptonemal complexes. In agreement with those observations, alcoholic PTA stains intensely the cores of the XY pair. Furthermore, optical observations with histochemical tests for DNA do not show DNA-containing condensations comparable to the cores; instead, "negative" filaments have been observed, which could be related to proteinaceous axes (Solari, 1964).

Thus, the present evidence supports the idea that the cores are not mainly formed by chromatin fibrils, but by some added material, part of which may be basic protein. However, this added material seems not to be unconnected to chromatin components; it is rather specifically deposited on some parts of the chromosomes at a specific moment, as Moses has suggested (1968). The autoassembly of a synaptonemal-complex-material (Schin, 1965), if it exists, would seem to be a rather rare event, and in any case, the involvement of extrachromosomal DNA in that assembly has not been discarded.

3. The Anomalous Synaptonemal Complex of the Long Core

If the core material is deposited at the interface between the two chromatids of the X chromosome, then the doubleness of this core may be related to an enlargement of this interface, and the deposition of that material on parallel sides of both chromatids. The organization of a synaptonemal complex at mid-pachynema would then be the result of the attainement of a "critical distance" of about 1,000 Å between two surfaces having homology, as suggested by Wolstenholme and Meyer (1966). However, the asymmetry of both lateral elements is a singular feature that is probably related with functional differences between sister chromatids. In relation with this asymmetry, Peacock (1965) has shown in *Plethodon cinereus* that the presumptive nucleolus-organizer region in the lampbrush chromosomes is asymmetric; one chromatid bears large loops related to nucleolus development and the other none.

In the present results it has been shown that the "redundant" or thicker lateral element of the anomalous synaptonemal complex was the one facing the developing granular zone of the nucleolus, with the intermediate chromatin which is probably connected to the thicker element. It can be suggested that the enlargement of chromatid interface and the excessive, asymmetrical deposition of synaptonemal material on one of the sides of this interface may be related to the synthesis of nucleolar components by one of the chromatids; when such a synthesis ends, as in late pachynema (complete development of the nucleolus), the anomalous synaptonemal complex becomes obliterated and the asymmetric array of the long core disappears. Experimental methods to test this hypothesis are available.

4. The Partial Synapsis of the X-Y Pair

The existence of synapsis in the X-Y pair of mammals has been doubtful according to most of the cytogenetical work (Sachs, 1955; Ohno, Kaplan and Kinosita, 1959). With the exception of a few species like the chinese hamster (Ohno and Weiler, 1962), there was no positive evidence either of synapsis inside the "sex vesicle" or of chiasmata in the sex bivalent. A hypothetical material of the "sex vesicle" was supposed to join both chromosomes by their ends.

The ultrastructural (Solari, 1964, 1969a) and histochemical (Solari and Tres, 1967a) analysis of the X-Y pair ("sex vesicle") showed the non-existence of materials other than the chromatin fibrils and the associated nucleolus in the case of the mouse [in which species the nucleolus organizer is carried by the X chromosome (Ohno, Kaplan and Kinosita, 1957)]. In the first ultrastructural study of the X-Y pair (Solari, 1964) no synaptonemal complexes were observed, mainly because full-size sex pairs during late pachynema were mainly studied (when the common end is obliterated) and no serial sections were used. However, the possibility of a *short, partial* synapsis was not excluded, and indeed it was soon observed that synaptonemal complexes were present in the X-Y pair (Solari and Tres, 1967b; Solari, 1969b). Ford and Woollam (1966) confirmed the ultrastructural description of the X-Y pair made by Solari (1964) and found a partial synaptonemal complex in the golden hamster.

As shown by the present results, the formation of a synaptonemal complex in the common end region between the two cores is a constant feature in every sex pair examined before late pachynema. A similar complex exists in man (Solari and Tres, 1969) and in the rat (in preparation), and probably in the golden hamster.

The restriction of the formation of the synaptonemal complex to the common end region may represent the lack of homology of the remaining part of the Y chromosome. The decrease in length of this synaptonemal complex during pachynema may be interpreted as a precocious repulsion of the X and Y chromosomes. However, the final $0.2\,\mu$ of this region remain joined up to diplonema. This fact is in agreement with cytological observations which show that the sex pair as soon as it is unravelled from the "sex vesicle" is already joined at the very tip of their ends. However, the functional significance of the apparent melting of the inner sides of the cores at the common end region must await further investigation.

Thus, lateral approachment and joining by the common end region has been proved in the X-Y pair, and synaptonemal complexes develop in that region as well as in paired autosomes. The presence of a synaptonemal complex is a prerequisite (Moses, 1968) for the formation of chiasmata but it is not in itself sufficient for that formation. Thus, although the existence of chiasmata and genetic recombination is suggested by the present results, further work is needed for its proof.

Acknowledgements. I thank Dr. L. Tres, Dr. R. Mancini and Miss M. Lema for help and support during this work. This work has been supported by a grant of the Population Council. The author is member of the Scientific Career, Consejo Nacional de Investigaciones.

References

Coleman, J. R., Moses, M. J.: DNA and the fine structure of synaptic chromosomes in the domestic rooster (*Gallus domesticus*). J. Cell Biol. 23, 63—78 (1964).

Ford, E. H., Woollam, D. H.: The fine structure of the sex vesicle and sex chromosome association in spermatocytes of mouse, golden hamster and field vole. J. Anat. (Lond.) 100, 787—799 (1966).

Meyer, G. F.: Possible correlation between submicroscopic structure of meiotic chromosomes and crossing-over. Third Europ. Conference on Electron Microscopy, B, 461—462 (1964).

Moens, P. B.: The structure and function of the synaptonemal complex in *Lilium longiflorum* sporocytes. Chromosoma (Berl.) 23, 418—451 (1968).

— The fine structure of meiotic chromosome pairing in the triploid *Lilium tigrinum*. J. Cell Biol. 40, 273—279 (1969).

Moses, M. J.: The relation between the axial complex of meiotic prophase chromosomes and chromosome pairing in a salamander (*Plethodon cinereus*). J. biophys. biochem. Cytol. 4, 633—638 (1958).

— Cytology and cell physiology (G. Bourne, ed.), p. 423—568. New York: Academic Press 1964.

Moses, M. J.: Synaptonemal complex. Ann. Rev. Genet. **2**, 363—412 (1968).

Ohno, S., Kaplan, W., Kinosita, R.: Heterochromatic regions and nucleolus organizers in chromosomes of the mouse, *Mus musculus*. Exp. Cell Res. **13**, 358—364 (1957).

— — — On the end-to-end association of the X and Y chromosomes of *Mus musculus*. Exp. Cell Res. **18**, 282—290 (1959).

—, Lyon, M. F.: Cytological study of Searle's X-autosome translocation in *Mus musculus*. Chromosoma (Berl.) **16**, 90—100 (1965).

— Weiler, C.: Relationship between large Y-chromosome and side-by-side pairing of the X-Y bivalent observed in the Chinese hamster, *Cricetus griseus*. Chromosoma (Berl.) **13**, 106—110 (1962).

Peacock, W. J.: Chromosome replication. In: Intern. Symp. on Genes and Chromosomes. Structure and Function. Nat. Cancer Inst. Monogr. **18**, 101—131. Bethesda: National Cancer Institut 1965.

Rhoades, M. M.: Meiosis. In: The cell (J. Brachet and A. Mirsky, ed.), vol. III, p. 1—75. New York: Academic Press 1961.

Reader, C., Solari, A. J.: The histology and cytology of the seminiferous epithelium of the mouse with Searle's X-autosome translocation. Acta physiol. lat.-amer. (in press).

Roth, T. F.: Changes in the synaptinemal complex during meiotic prophase in mosquito oocytes. Protoplasma **61**, 346—386 (1966).

Sachs, L.: The possibilities of crossing-over between the sex chromosomes of the house mouse. Genetica (s'Gravenhage) **27**, 309—322 (1955).

Schin, K. S.: Core-Strukturen in den meiotischen und post-meiotischen Kernen der Spermatogenese von *Gryllus domesticus*. Chromosoma (Berl.) **16**, 436—452 (1965).

Sheridan, W. F., Barrnett, R. J.: Cytochemical studies on chromosome ultrastructure. J. Ultrastruct. Res. **27**, 216—229 (1969).

Sjöstrand, F. J., In: Electron microscopy of cells and tissues. I. Instrumentation and techniques. New York: Academic Press 1967.

Solari, A. J.: The morphology and the ultrastructure of the sex vesicle in the mouse. Exp. Cell Res. **36**, 160—168 (1964).

— The evolution of the ultrastructure of the sex chromosomes (sex vesicle) during meiotic prophase in mouse spermatocytes. J. Ultrastruct. Res. **27**, 289—305 (1969a).

— Changes in the sex chromosomes during meiotic prophase in mouse spermatocytes. In: Nuclear physiology and differentiation. Genetics, Suppl. **61**, 113—120 (1969b).

—, Tres, L.: The localization of nucleic acids and the argentaffin substance in the sex vesicle of mouse spermatocytes. Exp. Cell Res. **47**, 86—96 (1967a).

— — The ultrastructure of the human sex vesicle. Chromosoma (Berl.) **22**, 16—31 (1967b).

— — The three-dimensional reconstruction of the X-Y pair of human spermatocytes. J. Cell Biol. (in press).

Sotelo, J. R., Wettstein, R.: Electron microscope study on meiosis. The sex chromosome in spermatocytes, spermatids and oocytes of *Gryllus argentinus*. Chromosoma (Berl.) **15**, 389—415 (1964).

Stern, H., Hotta, Y.: DNA synthesis in relation to chromosome pairing and chiasma formation. In: Nuclear physiology and differentiation. Genetics, Suppl. **61**, 27—39 (1969).

Wettstein, R., Sotelo, J. R.: Electron microscopical serial reconstruction of the spermatocyte I nuclei at pachytene. J. de Microsc. **6**, 557—576 (1967).

Wolstenholme, D. R., Meyer, G.: Some facts concerning the nature and formation of axial core structures in spermatids of *Gryllus domesticus*. Chromosoma (Berl.) 18, 272—286 (1966).

Woollam, D. H., Ford, E. H., Millen, J.: The attachment of pachytene chromosomes to the nuclear membrane in mammalian spermatocytes. Exp. Cell Res. 42, 657—661 (1966).

Distribution of chromosomes in metaphase plates of *Mesocricetus newtoni*

By PETRE RAICU, BARBU VLADESCU
and MARIA KIRILOVA

1. INTRODUCTION

While the relative constancy of the position of chromosomes in meiosis may not occur in somatic cell division, there is increasing evidence that a rather non-random distribution of the chromosomes in metaphase plates occurs during mitosis too. Schneiderman & Smith (1962) have shown that certain homologous chromosomes tend to lie together more frequently than would be expected by chance. Morishima, Grumbach & Taylor (1962) found that the late-replicating X chromosome displays rather peripheral locations in flattened metaphase figures, although German (1962) did not find any differences between the frequency of the peripheral location of the late-replicating X chromosome and that of the other chromosomes. Peripheral location of the Y chromosome in metaphase figures from cultured human leucocytes was reported by Miller *et al.* (1963 a) as well as specific location of some other chromosomes (Miller *et al.* 1963 b). The suggestion was made (Miller *et al.* 1963 a, b) that perhaps all the chromosomes tend to occupy specific positions.

The distribution of chromosomes in flattened metaphase spreads may reflect the distribution of chromosomes in the nuclear spindle equator, assuming that the colchicine and the hypotonic pretreatment do not have differential effects on specific chromosomes. If the chromosomes in the somatic cells undergo little relative movement during interphase, it may have a functional significance (Miller *et al.* 1963 a, b).

The present paper is an attempt to reveal any non-random distribution of chromosomes in male metaphase plates of the Rumanian hamster (*Mesocricetus newtoni*). The karyotype of *M. newtoni* consists of 18 pairs of chromosomes: 2 pairs of metacentrics, 5 pairs of submetacentrics and 11 pairs of subtelocentrics. The X chromosomes are the biggest subtelocentrics of the complement, while the Y is the smallest submetacentric (Raicu & Bratosin, 1966; Raicu, Hamar, Bratosin & Borsan, 1968).

2. MATERIALS AND METHODS

Male Rumanian hamsters from the Department of Genetics, University of Bucharest, were used in this study. Metaphase figures were obtained from bone marrow cells of animals previously injected with 0·06 % colchicine solution 2 h before killing. Hypotonic pretreatment was performed in sodium citrate, and fixation in a 3:1 mixture of alcohol and acetic acid. Aliquots of the suspension were dropped on clean slides and cells were quickly flattened and dried. The slides were stained in Giemsa solution and rinsed with water.

Metaphase figures were photographed and copies were made with a final magnification of × 2700. Only 51 metaphase plates with nearly equal diameters (about 60 mm) were selected to minimize somewhat the effect of the dispersion on the chromosome distribution.

The location of chromosomes was established by estimating the distance of the centromere of each chromosome from the centre of the metaphase plate, as determined from the mean of the co-ordinates of all the centromeres in the figure, and ascribing thus each chromosome to one of four equal concentric areas into which the metaphase figure was divided. The four equal concentric areas, designated as I, II, III, IV from the centre to the periphery, correspond each to 25 % of the total area of the metaphase figure.

The chromosomes in each metaphase figure were also classified as peripheral or non-peripheral in location by the method described by Miller *et al.* (1963a), and the results obtained by the two procedures were compared.

3. RESULTS AND DISCUSSION

The distribution of the chromosomes in the four equal concentric areas is shown in Table 1. Heterogeneity χ^2 calculated for the 80 (4 × 20) observations indicates a significant heterogeneity between chromosomes.

Some of the chromosomes in *Mesocricetus newtoni* metaphase figures appear to be distributed in a non-random fashion, as revealed by the χ^2 calculated for each chromosome, using the column totals to give expected values. Thus, X and Y chromosomes and the chromosome pairs 3 and 9 have a statistically significant tendency to occupy a peripheral position. On the other hand, chromosome pairs 16, 17 and 18 are located rather near the centre of the metaphase plate.

In comparing the distribution of chromosomes in the four areas with their peripheral or non-peripheral location, as established by the method of Miller *et al.* (1963a) (Table 2), some chromosomes which tend to lie in the outer part of the

Table 1. *Test of significance of the distribution of individual chromosomes**

(I, II, III, IV designate the four concentric areas into which
the metaphase plate was divided.)

Chromo-some	I	II	III	IV	Total	χ^2†	$P <$
X	6	13	10	22	51	16·86	0·01
Y	8	14	11	18	51	8·49	0·05
1	30	30	20	22	102	1·01	0·80
2	35	20	20	27	102	2·27	0·70
3	21	25	25	31	102	8·45	0·05
4	32	28	24	17	101	2·22	0·70
5	32	22	22	26	102	1·17	0·80
6	29	25	15	33	102	6·79	0·10
7	31	20	24	27	102	2·83	0·50
8	25	26	26	25	102	3·50	0·50
9	22	23	22	35	102	10·85	0·02
10	36	24	25	17	102	2·80	0·50
11	37	23	25	17	102	3·21	0·50
12	36	29	18	19	102	1·66	0·70
13	35	27	20	20	102	0·54	0·95
14	39	26	19	18	102	2·28	0·70
15	42	25	22	12	101	7·75	0·10
16	36	36	14	15	101	9·22	0·05
17	37	33	18	11	99	9·04	0·05
18	47	26	16	8	97	17·23	0·01
Total	616	495	396	420	1927	—	—

* Heterogeneity χ^2 for the 80 (4 × 20) observations = 118·5; D.F. = 57; $P < 0.01$.
† D.F. = 3.

Table 2. *Test of significance of the peripheral or non-peripheral location of individual chromosomes, as established by the method of* Miller et al. (1963a)

Chromosome	Total no.	Peripheral (%)	χ^{2*}	$P <$
X	51	50·98	14·08	0·01
Y	51	39·21	3·42	0·10
1	102	32·35	1·13	0·30
2	102	34·31	2·30	0·20
3	102	34·31	2·30	0·20
4	101	30·69	0·45	0·70
5	102	37·25	4·84	0·05
6	102	33·33	1·66	0·20
7	102	36·27	3·89	0·05
8	102	26·47	0·09	0·80
9	102	38·23	5·88	0·02
10	102	23·52	0·96	0·50
11	102	22·54	1·46	0·30
12	102	19·60	3·57	0·10
13	102	27·45	0·005	0·95
14	102	18·62	4·62	0·05
15	101	18·81	4·25	0·05
16	101	23·76	0·75	0·50
17	99	12·12	12·73	0·01
18	97	11·34	13·50	0·01
Total	1927	27·76	—	—

* With Yates' correction, D.F. = 1.

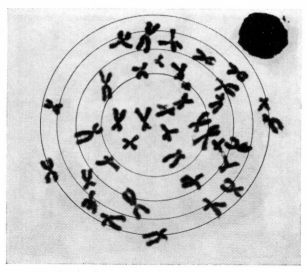

Fig. 1. The method used by us for the location of *Mesocricetus newtoni* chromosomes in four equal concentric areas.

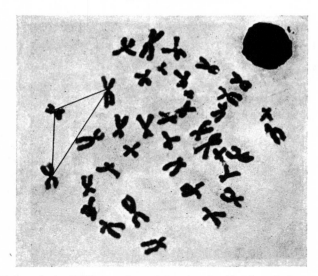

Fig. 2. The method of Miller *et al.* used for the classification of chromosomes as peripheral or non-peripheral. A chromosome is defined as peripheral if it lies outside the line connecting the peripheral chromosomes on either side of it.

metaphase plate (X, chromosome pair 9) appear to be peripheral according to Miller et al. too, and chromosomes scored as non-peripheral by this latter method appear to be located in the inner part of the plate (chromosome pairs 17 and 18).

There are, nevertheless, some discrepant cases, as the two methods do not always pick out the same chromosomes as peripheral or non-peripheral. It should be pointed out that a method based on counting the frequency of occurrence of chromosomes in four areas should be more reliable than a method such as Miller's, which only judges the position of each chromosome in regard to its neighbours.

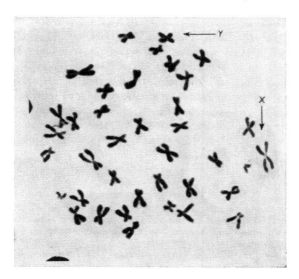

Fig. 3. The peripheral distribution of sex chromosomes in the metaphase plate of *Mesocricetus newtoni*.

It should be mentioned that the nomenclature in the *Mesocricetus newtoni* karyotype (Raicu & Bratosin, 1966; Raicu et al. 1968) is based only on the size of the chromosomes. The results reported here seem to suggest that the small chromosomes (16, 17, 18) are located at the inner part of the metaphase plate, while the big ones are rather peripheral in location. An exception is the Y chromosome, which, although rather small in size, is peripheral in location. The peripheral location of the Y chromosome suggests that the observed distribution of chromosomes is not a technical artifact, due to colchicine and hypotonic pretreatment, as chromosomes comparable in size with Y differ from it in location.

Miller et al. (1963b) suggested that the peripheral chromosomes are the late-replicating ones. This is probably true for the Y and X chromosomes in *Mesocricetus newtoni* too. As to the other chromosomes, no indication is yet available in this species.

The location of the X chromosome in flattened metaphase figures has not been established with certainty. However, there is some evidence supporting its peripheral location (Morishima *et al.* 1962; Grumbach, Morishima & Taylor, 1963; Miller *et al.* 1963 a, b). In our study, the highly significant peripheral location of the X chromosome in the metaphase figures of *Mesocricetus newtoni* is in good agreement with these observations.

As the inactivation of an X chromosome proceeds at random (Ohno & Cattanach, 1962; Lyon, 1963), it may be inferred that the male X corresponds to either of the two female X, and thus its location may suggest the peripheral location of the X chromosomes in both sexes. The tentative suggestion was made (Miller *et al.* 1963 a, b) that a correlation exists between the peripheral location of the X chromosomes and the sex chromatin, which is usually found at the periphery of the interphase nucleus in female mammalian cells (Barr, 1959).

If the peripheral location of chromosomes is dependent upon the later termination of replication, as heterochromatin terminates DNA replication later than euchromatin (Lima de Faria, 1961), it is likely that the functional differences which are responsible for the characteristic distribution of chromosomes may consist in their amount of heterochromatin (Miller *et al.* 1963 a, b).

The lack of information concerning DNA replication in *Mesocricetus newtoni* does not allow any conclusion about the connexion between the time of DNA replication and the location of chromosomes in this species.

REFERENCES

BARR, M. L. (1959). Sex chromatin and phenotype in man. *Science, N.Y.* **130**, 679–685.
GERMAN, J. L. (1962). Further characterization of the late X of the human female lymphocyte. In *Mammalian Cytology and Somatic Cell Genetics*. Conference held 15–17 Nov. 1962, Gatlinburg, Tennessee.
GRUMBACH, M. M., MORISHIMA, A. & TAYLOR, J. H. (1963). Human sex chromosome abnormalities in relation to DNA replication and heterochromatinization. *Proc. Natn. Acad. Sci. U.S.A.* **49**, 581–589.
LIMA DE FARIA, R. J. & BERGMAN, S. (1961). The pattern of DNA synthesis in the chromosomes of man. *Hereditas* **47**, 695–704.
LYON, M. F. (1963). Attempts to test the inactive-X theory of dosage compensation in mammals. *Genet. Res.* **4**, 93–103.
MILLER, O. J., MUKHERJEE, B. B., BREG, W. R., VAN, A. & GAMBLE, N. (1963a). Non-random distribution of chromosomes in metaphase figures from cultured human leucocytes. I. The peripheral location of the Y chromosome. *Cytogenetics* **2**, 1–14.
MILLER, O. J., BREG, W. R., MUKHERJEE, B. B., VAN, A., GAMBLE, N. & CHRISTOKOS, A. C. (1963b). Non-random distribution of chromosomes in metaphase figures from cultured human leucocytes. II. The peripheral location of chromosomes 13, 17–18 and 21. *Cytogenetics* **2**, 152–167.
MORISHIMA, A., GRUMBACH, M. M. & TAYLOR, J. H. (1962). Asynchronous duplication of human chromosomes and the origin of sex chromatin. *Proc. Natn. Acad. Sci. U.S.A.* **48**, 756–763.
OHNO, S. & CATTANACH, B. M. (1962). Cytological study of an X-autosome translocation in *Mus musculus*. *Cytogenetics* **1**, 129–140.
RAICU, P. & BRATOSIN, S. (1966). Le caryotype chez le *Mesocricetus newtoni*. *Z. Säugetierk.* **31**, 251–255.
RAICU, P., HAMAR, M., BRATOSIN, S. & BORSAN, I. (1968). Cytogenetical and biochemical researches in the Rumanian hamster (*Mesocricetus newtoni*). *Z. Säugetierk.* **33**,
SCHNEIDERMAN, L. J. & SMITH, C. A. B. (1962). Non-random distribution of certain homologous pairs of human chromosomes in metaphase. *Nature, Lond.* **195**, 1229–1230.

The Evolution of the Ultrastructure of the Sex Chromosomes (Sex Vesicle) during Meiotic Prophase in Mouse Spermatocytes

ALBERTO J. SOLARI

The structure and behavior of the X–Y pair during meiotic prophase in mammals have been a debated subject until recently. The debate over the existence of a "chromatin nucleolus" during male meiosis was settled by the work of several authors, specially Sachs (22) and Ohno (16, 17). Those authors established that the X–Y pair formed the so-called "sex vesicle" in spermatocytes. However, the identification of the sex vesicle at the ultrastructural level, made by Solari in 1964 (23) raised new questions. It was then assumed (23) that the granular component of the sex vesicle of mouse spermatocytes represented the main nucleolus of the spermatocyte. That assumption was proved to be true when the cytochemistry of the sex vesicle was studied and related to its ultrastructure by Solari and Tres (25). Further clarification of the ultrastructure of the sex vesicle was given by Solari and Tres (26) and by Ford and Woollam (2) in their studies on several species.

However, as meiotic prophase lasts about 10 days in the mouse, substantial changes occur in the sex vesicle during this period. Fortunately, the timing of each stage of meiotic prophase can be made accurately by the knowledge of the characteristic cellular associations of the spermatogenic cycle (15). The aim of this work is to describe the sequential changes in the ultrastructure and cytochemistry of the sex vesicle of mouse spermatocytes during meiotic prophase. The present results are followed by a discussion of the synaptic behavior of the X–Y pair and the development of the nucleolus in spermatocytes.

MATERIALS AND METHODS

Albino mice 4–10 weeks of age were used.

Smears for optical microscopy (25). Pieces of testes stripped from the albuginea were cut in isotonic saline at 4° and gently dispersed into isolated cells and cell aggregates by repeated aspiration in a 1-ml syringe. The suspension of cells was filtered through cloth and centrifuged, and the pellet was fixed with acetic acid–ethanol (1:3). Staining was done as previously described (25).

Thick sections. Sections 1 μ in thickness were cut in a Porter-Blum ultramicrotome from Maraglas embedded, glutaraldehyde-osmium tetroxide-fixed tissue, adhered to slides, and stained with polychrome blue.

Electron microscopy. Pieces of testes were fixed in 2.5% and 4% glutaraldehyde in 0.1 M phosphate buffer, pH 6.9, for 2 hours, washed in the same buffer plus 0.1 M sucrose, and fixed again in Caulfield's osmium fixative (1) for 1 hour. The pieces were embedded in Maraglas and sectioned in silver-colored sections. Serial sections were picked in single-hole grids (LKB Produkter) covered with Formvar films. Thicker sections (0.25 μ) were also used. Sections were stained in a saturated solution of uranyl acetate in methanol for 15 min, washed with methanol, and stained again with Reynolds' lead citrate (20). An Elmiskop I electron microscope was used.

Timing of the meiotic stages. The spermatogenic process is a very regular one in the mouse, and each cross section of a seminiferous tube can be classified into a particular stage of spermatogenesis by the characteristic cellular associations of each stage (15). To determine the stages of spermatogenesis, each piece of material was sequentially sectioned in thick (1 μ) and thin (silver-colored) sections. Thick sections, examined with the optical microscope, were labeled according to Oakberg's classification of the stages of spermatogenesis (15). A low magnification micrograph from the thin section was taken in the electron microscope to compare the optical and the electron microscopical pictures; then, the labeling of

---

FIG. 1. Resting spermatocytes, originated from type B spermatogonial mitosis. A cytoplasmic bridge joins two spermatocytes. Several dense masses of chromatin (arrows) are seen in their nuclei. × 5000.
FIG. 2. Spermatocytes at leptotene stage. The chromatin has become differentiated into patches of denser packing that may represent sections of thread-like condensations (arrow). × 8250.
FIG. 3. Nucleus of one spermatocyte at zygotene (stage 11 of spermatogenesis). The sex vesicle (*sv*) forms a lighter mass of chromatin in which cores (arrows) are seen. Synaptinemal complexes (left arrow) are seen in the autosomal pairs, one of them ending on a dense, peripheral chromatin mass. × 15,000.

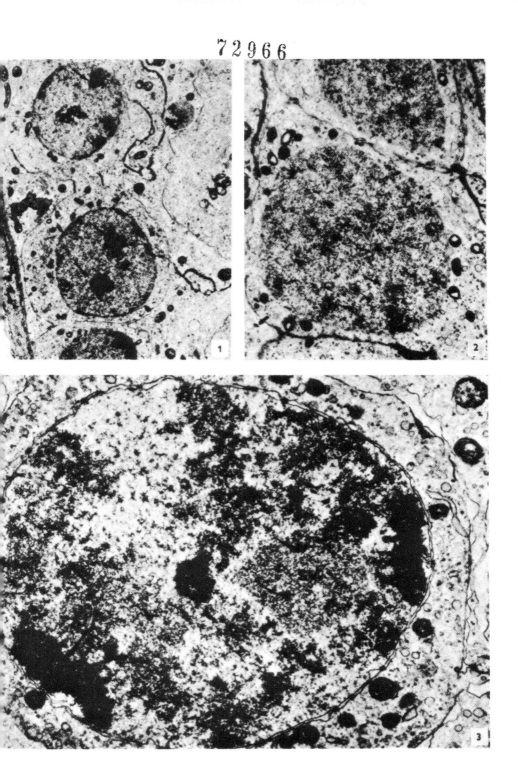

the stage was made in the electron micrograph. The smears used for the histochemical observations could not be labeled so accurately as the sections. The time sequence of the spermatocyte pictures in the smears was based on the degree of chromosome condensation and pairing, and on the development of the nucleolus. A good correspondence was generally obtained between the time sequence of sections and smears, but a few cells in the smears could not be classified.

The long pachytene stage was divided into early pachytene (stages 1–3), middle pachytene (stages 4–7), and late pachytene (stages 8–10) in order to describe the ordered changes of the sex vesicle.

RESULTS

The sex chromosomes in resting spermatocytes and leptotene
(stages 7–9, first layer of spermatocytes)

Resting spermatocytes (defined as the earlier stage after the division of type B spermatogonia) form rows of several cells at the same stage (Fig. 1). They are separated from the base membrane by Sertoli cell expansions. The nuclei of the resting spermatocytes have a few large, oblong masses of condensed chromatin that may appear round in thin sections (Fig. 1). Those chromatin masses are in contact with the nuclear membrane and their longer diameters are radially arranged. The evolution of those chromatin masses proved that they do not represent the sex chromosomes (see below). The remaining chromatin of the resting spermatocytes is rather diffuse and homogeneous (Fig. 1). Thus, no identification of the sex chromosomes in thin sections is possible at this stage.

Leptotene nuclei are characterized by the beginning of the condensation of the chromatin, which forms diffuse thin threads (Fig. 2). Although at the center of the threads there are axial condensations, they are not so sharply delimited and stained as the cores of later stages. The large blocks of condensed chromatin seen in the resting spermatocytes are now somewhat flattened against the nuclear membrane and show axial condensations about 300 Å in diameter. Occasionally, masses of chromatin fibrils having a diffuse appearance and without cores were seen in those nuclei. The possibility that they represent the sex chromosomes could not be proved, although those chromatin regions have the same degree of condensation characteristic of the sex vesicle at zygotene.

The sex chromosomes at zygotene
(stages 11–12, first layer of spermatocytes)

The spermatocyte nuclei at zygotene are characterized by two sharp features: (a) the formation of abundant synaptinemal complexes associated with chromatin

---

FIG. 4. Higher magnification of a sex vesicle at zygotene. The sex vesicle has a dense, thick core (arrow). The sex vesicle is not yet connected to the nuclear envelope (right arrow). × 39,600.

FIG. 5. Spermatocyte nucleus at early pachytene. The sex vesicle is already broadly associated to the nuclear membrane and has developed a clear space around its inner side. × 16,500.

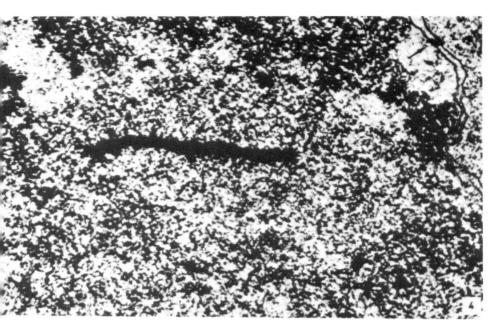

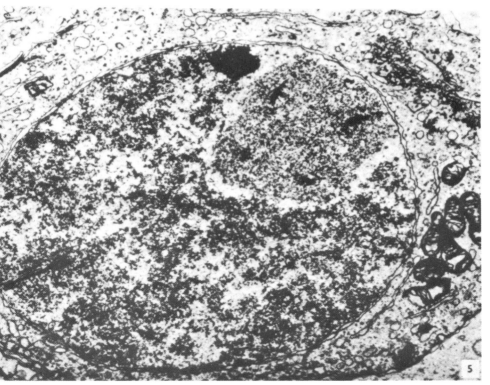

threads, and the observation of single cores in other chromatin threads; and (b) the formation of the sex vesicle (Fig. 3). The sex vesicle is formed by a mass of chromatin fibrils having a characteristic and homogeneous degree of condensation (Fig. 3). The degree of condensation of the sex vesicle in this stage is weaker than that of the autosomes. The sex vesicle is not yet broadly connected to the nuclear membrane, and it is centrally located in some sections; it appears as a light, oblong mass up to 3 μ long inside the nucleus (Fig. 3). Sometimes two similar, unconnected masses are seen. The sex vesicle has thick cores that are different from the synaptinemal complexes that are in the remainder chromatin. The sections of the cores of the sex vesicle (Figs. 3 and 4) show that they are rather straight at this stage, having lengths up to 1.5 μ in thin sections (Fig. 4). The cores are formed by two tightly joined filaments, each one 300–400 Å wide, forming a couple 900–1100 Å wide (Figs. 4 and 9). Both filaments seem to be plectonemically coiled (Fig. 4). The chromatin fibrils that form the sex vesicle are 120–150 Å at their thicker points and 30–40 Å at the thinner ones (Fig. 6).

Two kinds of ends of the sex vesicle cores can be seen. Most of the ends are those in which the two 300 Å filaments end separately on the nuclear membrane, but without forming a synaptinemal complex. In a very few instances, a synaptinemal complex was seen inside the vesicle ending on the nuclear membrane; in those cases, both lateral components diverged to opposite directions after a short path.

No nucleolar part is yet developed in the sex vesicle at zygotene. Some small, dense, ribbon-like elements of leptotene nuclei have now developed into a small peripheral region of granular structure; they are interpreted as secondary nucleoli, independent from the sex vesicle.

The synaptinemal complexes seen in the remainder chromatin, are 2000 Å wide in cross and frontal sections (see Fig. 6). The lateral components are 400–500 Å wide in frontal sections, and they are separated by a space 1000 Å wide. The central component is 150 Å wide in frontal sections (Fig. 6). The synaptinemal complexes do not differ in dimension from those of early and middle pachytene.

The sex chromosomes at early pachytene
(stages 1–3, spermatocyte layer)

The condensation of autosomal chromatin increases up to early pachytene, and from then on a slow uncondensation takes place up to diplotene. Thus, at early pachy-

---

FIG. 6. Small part of a nucleus at early pachytene. A part of a sex vesicle is at left, showing its cores (arrows). At right, a piece of autosomal chromatin shows a synaptinemal complex running toward the nuclear membrane (right arrow). × 55,800.

FIG. 7. Part of a spermatocyte nucleus at middle pachytene (stage 6 of spermatogenesis). The sex vesicle pushes the nuclear membrane and forms a protruding mass. A core ends as a structure identical to a synaptinemal complex (arrow). × 21,000.

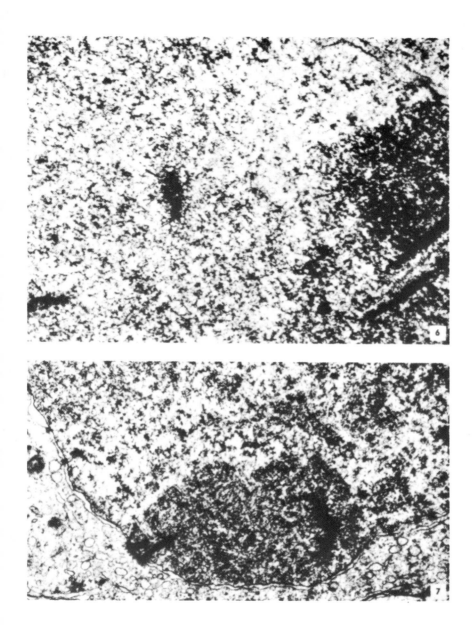

tene, the autosomes have synaptinemal complexes surrounded by rather condensed chromatin, and the sex vesicle stands out lighter and more homogeneous than the autosomes (Fig. 5). The sex vesicle is already broadly contacting with the nuclear envelope, and it has developed a light area around it that marks a sharp boundary to the inner part of the sex vesicle (Fig. 5).

No nucleolar region has yet been developed in the sex vesicle, although in some sections, a dense, round structure, 0.5 μ wide is found near the inner side of the vesicle. That round body is probably the structure found inside the nucleolar region of the sex vesicle at later stages.

The cores inside the sex vesicle are similar to those found in zygotene except that they are more flexuous and that the separation of both filaments of the core is more clearly seen (Fig. 6). A comparison between the sex vesicle and an autosomic synaptinemal complex is seen in Fig. 6.

The sex chromosomes at middle pachytene
(stages 4–7, spermatocyte and upper spermatocyte layers)

At middle pachytene, specially at stage 6, synaptinemal complexes are most easily seen inside the less condensed chromatin. Most synaptinemal complexes are seen cut in cross or oblique sections. The autosomes have now a similar density compared with the sex vesicle (Fig. 7). The sex vesicle remains clearly delimited by the light and narrow space on its inner side (Fig. 7). The contact between the sex vesicle and the nuclear membrane is so well developed that sometimes the sex vesicle forms a protrusion toward the cytoplasm (Fig. 7). A dense, round body becomes associated to the inner side of the sex vesicle. That round body is surrounded by a narrow layer of granules 200 Å wide. The higher frequency of synaptinemal complexes inside the sex vesicle was found at this stage. Thus, the three-dimensional reconstruction of the sex vesicle was made at this stage. The results of that reconstruction are described elsewhere (*24*), but a summary will be given here. Two independent cores run inside the sex vesicle. The short one has the same characteristics as the cores described in zygotene. The longer core, however, has developed a structure identical to a synaptinemal complex at some parts of its length (Fig. 7). Three endings of the cores on the nuclear membrane are seen: one end in which two extremes (one from each core) are associated, one free end of the short core and one free end of the long core. The free end of

---

FIG. 8. Nucleus of a spermatocyte at very late pachytene (stage 10 of spermatogenesis). The section passes through the thicker part of the nucleolar region of the sex vesicle, which has reached its maximum development. Most of the autosomal chromatin is uncondensed.   12,000.

FIG. 9. Part of a spermatocyte nucleus at diplotene (stage 11). The chromatin part of the sex vesicle is separated by a narrow space (arrow) from the nucleolar region. Double, dense cores remain in the sex vesicle. × 20,000.

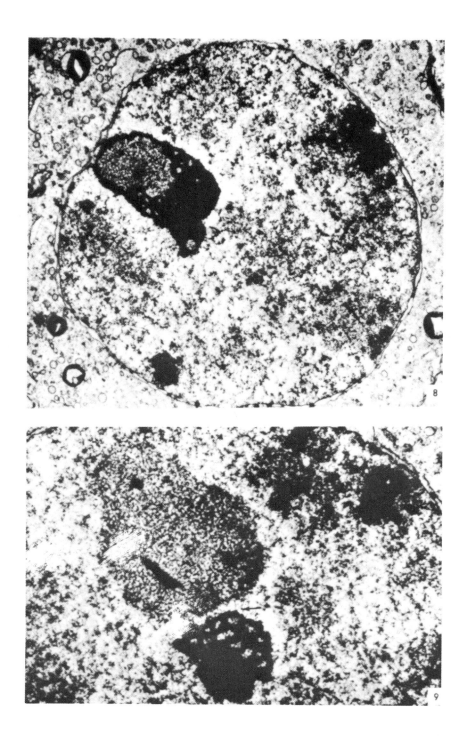

the long core is identical to a synaptinemal complex. The structure of the common ending is complex, and there is probably a connection between the cores (24).

The sex chromosomes at late pachytene
(stages 8, 9, and 10, upper layer of spermatocytes) (Figs. 8 and 10)

The decondensation of the autosomes comes to a maximum at this stage; thus, by contrast, the sex vesicle is comparatively darker than the autosomes. Inside the sex vesicle the cores no longer show synaptinemal complexes. Endings of the cores are composed of two filaments tightly joined, each one being 400–500 Å wide. Frequently cross sections of cores have an oval profile. The nucleolar region develops strongly during this period (Figs. 8 and 10). A cup-like layer of granular material covers most of the inner surface of the sex vesicle. When fully developed, the nucleolar region of the sex vesicle consists of several parts (Fig. 10): (a) the granular layer, composed of granules about 200 Å wide and a less visible fibrillar component; this layer has a half-moon shape, being thicker in the part which surrounds the round body; (b) invaginations of the granular layer into the chromatin part of the sex vesicle; those invaginations are thread-like, about 1000 Å wide and penetrate deeply into the chromatin, but they do not touch the cores; the invaginations come from both the central part of the half-moon and its ends; (c) a dense, round body in which the 200 Å granules are much more tightly packed; it is 0.5–0.9 μ in diameter; (d) an oblong, dense body, about 1 μ in length, of predominantly fibrillar structure; this oblong body protrudes from the half-moon toward the nuclear sap, contains a round, less condensed area, and frequently has a ring-like body at one end. The round body and the oblong body are distinguished from the half-moon by their histochemical properties.

The sex chromosomes at diplotene and diakinesis
(stages 11–12, upper spermatocyte layer)

The sex vesicle adopts a special packing during this stage (Fig. 9). The average thickness of the chromatin fibrils seems to be larger than before, but regular spacings, about 300 Å, remain, separating the chromatin fibrils (or threads) (Fig. 9). Sections of cores, formed by two filaments, stain strongly (Fig. 9). The nucleolar part becomes detached from the sex vesicle at some stage during diplotene (Fig. 9); no nucleolar invaginations remain inside the sex vesicle. The nucleolar region diminishes rapidly in volume during diplotene and diakinesis (Fig. 11) and becomes a dense mesh in which the 200 Å granules are less prominent (Fig. 11).

FIG. 10. Higher magnification of the sex vesicle from the same cell as Fig. 8 (late pachytene). The granular layer (*gr*) surrounds completely the chromatin part of the sex vesicle because of the section incidence. The round body (*rb*) is surrounded by the granular layer. The oblong body (*ob*) forms a protruding mass ending on a ring-like profile, and it has a round, lighter part at the other end. × 53,000.

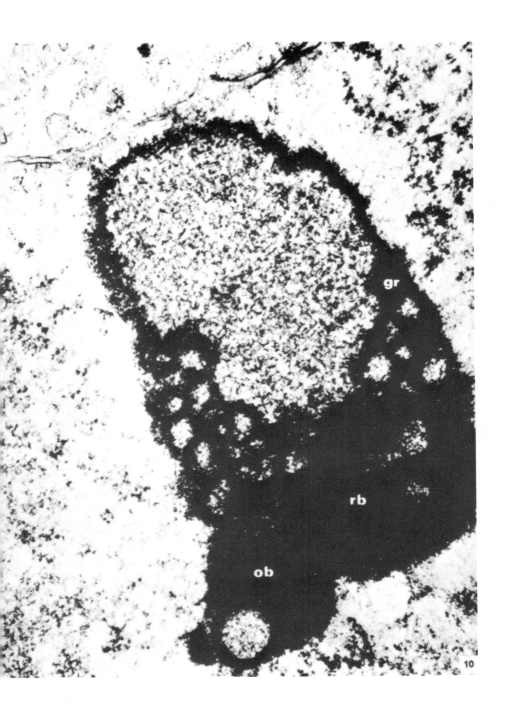

The sex chromosomes after diakinesis (stage 12)

Fifty cells were recorded at the metaphase, anaphase, and telophase stages of the first meiotic division. No special condensation of any chromosome was observed in those cells. As the complete shape of each bivalent cannot be seen in thin sections, the identification of the X–Y pair was not possible. However, a chromosome was seen having a sharp, dense core at those stages (Figs. 12 and 13). The core has a similar width to those of the sex vesicle at diplotene, and it has the same structure. It runs near the centromere region (Fig. 13), but inside the chromosomal mass.

As the staining affinity and the thickness of the cores of the sex vesicle at diplotene are larger than the remainder cores of the autosomes, the metaphase-anaphase core probably belongs to one of the sex chromosomes.

The cytochemical changes of the sex vesicle

The cytochemical characteristics of the late pachytene sex vesicle have been described earlier (25). The examinations of the smears stained with the Feulgen, the methyl-green and pyronin, and the acridine-orange fluorochromation techniques, showed the changes in DNA- and RNA-containing structures. Leptotene and zygotene nuclei do not possess large or median nucleoli; only some very small and scattered masses give the RNA reactions. The sex chromosomes are first identified at zygotene, either joined or separated, by their shortened length and homogeneous reaction with DNA tests. No RNA is detected in the sex vesicle at that stage.

Secondary nucleoli are seen in one or two autosomes at early pachytene. The sex vesicle does not show the presence of RNA, but axial condensations are seen with the DNA tests. At middle pachytene some RNA-containing material is found on one of the sides of the vesicle, while secondary nucleoli attain their maximum development in the autosomes. At late pachytene, a well developed, half-moon region containing RNA covers about one-half of the sex vesicle. Inside the half-moon, a round space is negative for RNA reactions, corresponding in size and location with the "round body" described earlier. RNA invaginations into the chromatin part of the sex vesicle are seen, especially with acridine-orange fluorochromation.

At diplotene-diakinesis, the RNA-containing region is separated or only touches the sex vesicle.

FIG. 11. Nucleus of a spermatocyte at diakinesis (stage 12). The nucleolus forms a loose mesh; its relationship to the sex chromosomes cannot be decided in this case. Autosomes have strongly condensed against the nuclear membrane. $\times$ 12,500.

FIG. 12. First meiotic anaphase showing a curved, dense core in a chromosome. $\times$ 16,000.

FIG. 13. First meiotic telophase nucleus showing the dense core (arrow), and besides a double centromeric region. $\times$ 18,000.

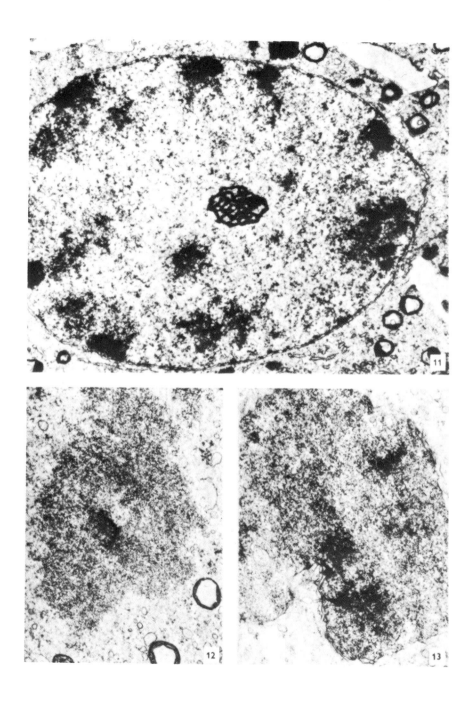

DISCUSSION

The first description of an heterochromatic body in the nuclei of mammal spermatocytes was probably made by Lenhossek in the last century (3). This body was described later by other cytologists, specially Regaud (19), but its identity with the sex chromosomes was not noticed at that time. The nature of this heterochromatic body was clarified by the works of Painter (18), Minouchi (7), and Makino (4) on several mammals, and especially by Sachs (22) in man and by Ohno (16, 17) in the mouse and the rat. The latter studies established that that body is formed by the X and Y chromosomes at zygotene and that it lasts up to diplotene, when in most mammals the X and Y chromosomes appear connected end-to-end (17). The name "sex vesicle," given by Sachs (22) to that structure, rests on the assumption that a special material is embedding the sex chromosomes at those stages. However, as Solari and Tres (25) and Solari (23) proved that there is neither a special material nor a membrane in the sex chromosomes, the "sex vesicle" should be more correctly labeled as the *heterochromatic sex pair*.

The chromatin nature of the sex vesicle was obscured by some exceptional cases, such as the mouse, in which the sex chromosomes bear the nucleolus organizer; thus, in the mouse the main nucleolus is attached to the sex vesicle. However, as shown earlier (25) and in the present work, the nucleolar and the chromatin parts of the sex vesicle of the mouse can be delimited, and besides, the nucleolar part does not exist or it is detached from the chromatin part at some stages of meiotic prophase. Studies on other mammals (2, 26) have shown that the sex vesicle is made essentially by chromatin fibrils, and that it is comparable to the outer (fibrillar) part of the sex vesicle of the mouse. Thus, the sex vesicle in most mammals is composed by the X and Y chromosomes differentially condensed and attached to the nuclear membrane.

The development of the main nucleolus in the sex vesicle of the mouse

The sex vesicle does not show granular elements during zygotene and early pachytene. However, at early pachytene the small round body composed by dense fibers can become connected to the inner side of the sex vesicle. The behaviour of that round body, that is, its covering by granular elements and its own granular structure at later stages, suggests that this round body is the first nucleolar structure associated with the sex vesicle. The inability to demonstrate RNA in this round body suggests that this body may be similar to the "nucleoli" of fertilized eggs in mammals, which are fibrillar bodies that contain little or no RNA (29).

The granular part that develops from middle to late pachytene shows the characteristic RNA reactions. It has a structure very similar to that of the main part of the independent main nucleolus in human spermatocytes at pachytene (30). Human (and rat) nucleoli have three zones or layers: the main one, peripheral and formed by granules; a central one, that has a fibrillar, dense structure; and an intermediate part, less dense and connected to chromatin fibrils. It is tentatively suggested that the granular layer, the oblong body, and the round body described in the nucleolar part of the sex vesicle at late pachytene are homologous to the three zones found in the human nucleoli.

The sex chromosome of *Gryllus argentinus* spermatocytes becomes associated with the nucleolus during meiotic prophase (28). However, that association seems not to be comparable with the structure found in the mouse. Wettstein and Sotelo (31) have described a different three-dimensional pattern and a different evolution in the sex chromosome-nucleolus association in *Gryllus*.

The differential condensation of the chromatin in the sex vesicle

The chromatin of the sex vesicle shows approximately the same degree of condensation throughout the meiotic prophase, sharply contrasting with the changes in the autosomal chromatin. The autosomes go through an uncondensation process at middle and late pachytene, and at that time their metabolic activity is rather high (9). The lack of that chromatin loosening in the sex vesicle agrees with its lower metabolic activity (9, 10).

It must be remarked that the packing of the sex vesicle chromatin is not so strong as that of the peripheral, heterochromatic blocks of some autosomes. Regular spacings are conserved among the chromatin fibrils of the sex vesicle through most of the stages, and diplotene condensation is made rather by the thickening of chromatin fibers than by a more close packing of the fibers. The sex chromatin (Barr's body) also shows a peculiar packing of the chromatin fibers (32).

The evolution of the cores in the sex vesicle

Solari (23) observed that no synaptinemal complexes were present in the sex vesicle while describing for the first time its ultrastructure. However, as seriated sections were not made at that time, the existence of a short synaptinemal complex was explicitly not exluded (23). The present study, by a systematic screening of spermatocytes at every stage of meiotic prophase has shown more complex facts. No synaptinemal complexes are found in the sex vesicle at late pachytene, and very few were seen at early pachytene. However, at middle pachytene, some parts of the cores form a structure identical to a synaptinemal complex. As it has been shown by serial sectioning that that structure is formed at some places along the long core (24), its meaning is probably different from that of the synaptinemal complexes found in the autosomes; that is, it cannot represent the synapsis of two homologs. It is assumed tentatively that that "synaptinemal complex" is evolved at some moment of the prophase from the two filaments that form the long core, and that afterward, at late pachytene, it is obliterated. The oval profiles of the sectioned cores at late pachytene suggest that a tight relational coiling develops from both filaments of the core. At late pachytene and diplotene a partial disintegration of the cores seems to occur. However, no "stripped cores" (8) have been seen connected with the sex vesicle. A part of the cores remains in the sex vesicle up to the latest stages of prophase, when it is difficult to see any core in the autosomes. The permanence of the core in one of the sex chromosomes during metaphase through telophase may be related to the special condensation of that chromosome. The meaning of the synaptinemal complex in meiotic prophase has been extensively studied by Moses (11–13) and by other authors (8, 14, 21, 27). Moses' demonstration of the relationship between the pairing of homologs and the synaptinemal complex has been confirmed in most cases (6, 8, 14, 21). However, Sotelo and Trujillo-Cenoz (27) and Moses (13) first noticed exceptional cases in which similar structures could not be easily explained by a pairing of homologs. Sotelo and Wettstein (28) and Wettstein and Sotelo (31) have shown that synaptinemal complexes are involved in the structure of the single X chromosome in *Gryllus argentinus*. Further cases have been described (5, 21) of atypical complexes. The present observations in the sex vesicle show another case in which probably that complex does not represent the synapsis of homologous chromosomes.

However, a region of association between the short and the long cores exists in the sex vesicle of the mouse (24). It has been proved by the use of serial sections (24) that a partial, short synaptinemal complex is formed in the common end region of both cores at early pachytene. Thus, a short homologous region exists in the X and Y chromosomes of the mouse. This subject is further discussed by Solari (24).

The collaboration of Dr. L. Tres is gratefully acknowledged. This work has been supported by a grant from the Consejo Nacional de Investigaciones and a grant from the Population Council (C.M. 6713) to Professor R. E. Mancini, whose support is thankfully acknowledged. The author is an Established Investigator from the Consejo Nacional de Investigaciones Científicas.

REFERENCES

1. CAULFIELD, J. B., *J. Biophys. Biochem. Cytol.* 3, 827 (1957).
2. FORD, E. H. and WOOLLAM, D. H., *J. Anat.* 100, 787 (1966).
3. LENHOSSEK, M. v., *Arch. Mikroskop. Anat.* 51, 215 (1898).
4. MAKINO, S., *J. Fac. Sci. Hokkaido Univ. Ser. VI* 7, 305 (1941).
5. MENZEL, M. Y. and PRICE, J. M., *Am. J. Botany* 53, 1079 (1966).
6. MEYER, G. F., *Proc. European Regional Conf. Electron Microscopy, Delft*, Vol. 2, 951 (1960).
7. MINOUCHI, O., *Japan. J. Zool.* 1, 235 (1928).
8. MOENS, P. B., *Chromosoma* 23, 418 (1968).

9. MONESI, V., *Exptl. Cell Res.* **39**, 197 (1965).
10. —— *Chromosoma* **17**, 11 (1965).
11. MOSES, M. J., *J. Biophys. Biochem. Cytol.* **2**, 215 (1956).
12. —— *ibid.* **4**, 633 (1958).
13. MOSES, M. J. and COLEMAN, J.,R., *in* LOCKE, M. (Ed.), The Role of Chromosomes in Development, p. 11. Academic Press, New York, 1964.
14. NEBEL, B. R. and COULON, E. M., *Chromosoma* **13**, 272 (1962).
15. OAKBERG, E. F., *Am. J. Anat.* **99**, 391 (1956).
16. OHNO, S., KAPLAN, W. and KINOSITA, R., *Exptl. Cell Res.* **11**, 520 (1956).
17. —— *ibid.* **18**, 282 (1959).
18. PAINTER, T. S., *J. Exptl. Zool.* **39**, 197 (1924).
19. REGAUD, C., *Arch. Anat. Microscop.* **11**, 291 (1910).
20. REYNOLDS, E. S., *J. Cell Biol.* **17**, 208 (1963).
21. ROTH, T. F., *Protoplasma* **61**, 346 (1966).
22. SACHS, L., *Ann. Eugenics* **18**, 242 (1954).
23. SOLARI, A. J., *Exptl. Cell Res.* **36**, 160 (1964).
24. —— *Intern. Symp. Nuclear Physiol. Differentiation, Belo Horizonte, 1968, Genetics* (in press).
25. SOLARI, A. J. and TRES, L., *Exptl. Cell Res.* **47**, 86 (1967).
26. —— *Chromosoma* **22**, 16 (1967).
27. SOTELO, J. R. and TRUJILLO-CENOZ, O., *Z. Zellforsch. Mikroskop. Anat.* **51**, 243 (1960).
28. SOTELO, J. R. and WETTSTEIN, R., *Chromosoma* **15**, 389 (1964).
29. SZOLLOSI, D., *J. Cell Biol.* **25**, 545 (1965).
30. TRES, L. and SOLARI, A. J. (in preparation).
31. WETTSTEIN, R. and SOTELO, J. R., *J. Microscopie* **6**, 557 (1967).
32. WOLSTENHOLME, D. R., *Chromosoma* **16**, 453 (1965).

CHANGES IN THE SEX CHROMOSOMES DURING MEIOTIC PROPHASE IN MOUSE SPERMATOCYTES

ALBERTO J. SOLARI

THE X-Y chromosome pair of mammals during meiotic prophase forms a prominent heteropycnotic body which, although it has received a variety of names (MITTWOCH, 1967), is most commonly referred to as the "sex vesicle" (SACHS, 1954). The fine structure, histochemistry and behaviour of the sex pair has been investigated in our laboratory during the past five years (SOLARI, 1964; SOLARI and TRES, 1967a; SOLARI and TRES, 1967b; TRES and SOLARI, 1968; SOLARI. 1968; READER and SOLARI, 1968). The aim of this paper is to review the previous results on the structure of the sex vesicle in the mouse and to present new evidence on the nature of synapsis between the X and Y chromosomes.

MATERIALS AND METHODS

Randomly bred albino mice were most commonly used in these investigations. Mice carrying Searle's X-autosome translocation (T(X; ?)16 H) were generously supplied by Dr. B. M. Cattanach (City of Hope Medical Center, Duarte, Cal.). Histochemical and electron microscopical procedures have been previously described (SOLARI and TRES 1967a and 1967b). The sequential ordering of meiotic stages was based on OAKBERG's (1956) description of the spermatogenic cycle of the mouse. Three-dimensional models of the ultrastructure of the sex pair were made with celluloid sheets at scale, based on seriated sectioning of complete sex pairs.

RESULTS

Formation and changes of the sex pair (sex vesicle) during meiosis: Permanent heteropycnosis of the sex chromosomes is not observed before the spermatocytic stage. The sex chromosomes become heteropycnotic at leptotene; in smears and squashes treated with acridine-orange they appear as chromocenters having the fluorescence typical of DNA, but they show no RNA reactions, nor are they related to structures having RNA at leptotene. The sex pair becomes a single and characteristic structure ("sex vesicle") at zygotene. In electromicrographs it appears to be a mass of chromatin fibrils having a characteristic and homogeneous degree of condensation (SOLARI, 1964, SOLARI and TRES, 1967a). The sex pair is about 3 microns in length and it is not yet broadly connected to the nuclear membrane. Unpaired cores, about 1,000 Å wide and composed of two tightly coiled filaments about 400 Å wide, are present in the sex pair at zygotene, while in the autosomes the characteristic synaptinemal complexes are seen. The histochemical tests, at both the optical and electronmicroscopical levels, do not show RNA in the sex pair at zygotene (SOLARI and TRES, 1967a).

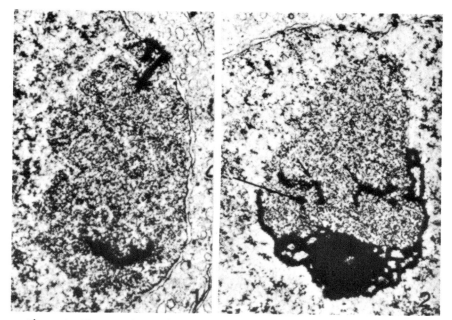

FIGURE 1.—Electron micrograph of the sex pair at middle pachytene, a dense mass contacting the nuclear membrane. A synaptinemal complex ends against the membrane. × 21,000.

FIGURE 2.—Electron micrograph of the sex pair at late pachytene. The nucleolar part consists of a granular layer, a dense round body and granular invaginations inside the chromatin (short arrow). Two sections of cores (long arrow) are seen. × 18,000.

The 10-day long pachytene stage may be divided into early, middle and late pachytene (SOLARI, 1968). The sex pair becomes broadly attached to the nuclear membrane at early pachytene. The cores are similar to those of the sex pair at zygotene. The three-dimensional reconstructions show that the cores are unpaired except at one special place, the *common end* (see below). A clear halo separates the inner part of the sex pair from the remaining nuclear structures.

During middle pachytene an important change occurs in the long core. This core shows at some places a structure identical to a synaptinemal complex; it is the *atypical synaptinemal complex* (Figure 1). The common end region of both cores is present but shortened, compared to early pachytene. A small round body, 0.3–0.4 micron in diameter, is visible at the inner side of the sex pair. A narrow layer of dense granules, about 200 Å each, has developed around the round body and is beginning to cover the inner side of the sex pair. Histochemical tests at optical and electronmicroscopical levels show that the granular layer contains RNA, but that the round body does not give positive reactions for RNA.

During late pachytene the sex pair attains its maximum development by virtue of the growth of the nucleolar region of the sex pair which consists of several parts (Figure 2): a) a granular layer, formed by 200 Å wide granules, extending as a half-moon covering most of the inner side of the sex pair; b) a dense, round body centered at the granular layer; c) an oblong, fibrillar body; and d) invaginations of the granular layer in the chromatin mass. The atypical synaptinemal complexes have disappeared at this stage, although the long core shows a bipartite structure.

At diplotene and diakinesis the nucleolar region becomes separated from the chromatin region and diminishes in volume. The fate of the sex chromosomes after diakinesis is discussed by SOLARI (1968).

Three-dimensional arrangement of the axial cores of the sex chromosomes: The spatial reconstruction of the sex pair by models based on seriated sections was done at early, middle and late pachytene. In every case two cores with the same characteristics were present. Furthermore, these cores are rather uniform in length, the long core averaging 8.4 microns and the short one 3.4 microns. Thus, the two cores are permanent and characteristic structures and will be called long and short cores (Figure 3). At very early pachytene the long and the short cores are axial to oblong, sharply delimited masses of chromatin that merge at one side, the long core being axial to the larger mass. At middle and late pachytene both cores are inside the common mass of chromatin of the sex pair.

Two kinds of ends of the cores are always observed in each sex pair: the *common end* and the *free ends*. The long and the short cores come near and parallel to each other at one end, the common end, where they finally touch the nuclear membrane. Each core also has one free end (Figure 3).

The common end segment of the cores varies in length, being maximum at early pachytene (0.95 micron) and diminishing up to late pachytene (0.2 micron or less). This region, when seen at early pachytene, although having the same structure as a synaptinemal complex, differs in that the lateral components of the complex separate and take opposite directions after the short synaptic region thus forming the long and the short cores (Figure 3). The synaptic region will be referred to as the *typical* synaptinemal complex (see discussion). The synaptic region is almost null at medium and late pachytene; the lateral components of the synaptinemal complex are present but shortened and the central component is either absent or replaced by 2 or 1 transverse or oblique filaments about 200–300 Å wide that join both lateral components. The free end of the long core is surrounded by a differentially condensed chromatin, similar to that of the autosomal ends, at early pachytene. The path of the short core is simple because it mostly lies in a single plane; but the long core is progressively convoluted from early to late pachytene. The long core shows a bipartite structure at early and late pachytene, but at middle pachytene it has at many places a structure identical to the atypical synaptinemal complex, because it is formed only by the long core. The ratio between the lengths of the long and short cores is uniform, about 2.5.

Comparative aspects of the sex pair in mammals: The human sex vesicle lacks the nucleolar region and has much longer and convoluted cores but other char-

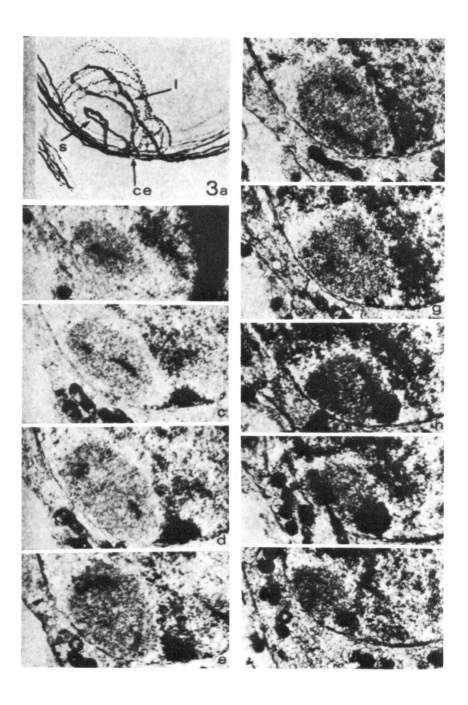

acteristics are similar to that of the mouse sex vesicle (SOLARI and TRES, 1967b). The sex pair of the rat, the guinea pig and the Chinese hamster also lack a nucleolar region and show nucleoli independent from the sex pair.

Metabolic activities in the sex pair at prophase: The uptake of tritiated uridine and tritiated cytidine given by intratesticular injection was studied by autoradiographical techniques. The sex pair does not show any significant uptake at any stage of meiotic prophase, thus confirming MONESI's (1965) observation.

The sex pair in Searle's translocation (SEARLE, 1962; OHNO and LYON, 1965): The sex pair in mice bearing Searle's translocation shows the following features when studied with optical and electronmicroscopical methods (READER and SOLARI, 1968, SOLARI, in preparation): 1) The nucleolar region is diminished in size; 2) the size of the sex vesicle is enlarged; 3) there is a connection between the sex vesicle and two different masses of chromatin having autosomal condensation; and 4) there is physical continuity between the lateral components of autosomal synaptinemal complexes and the long core of the sex vesicle.

DISCUSSION

OHNO, KAPLAN and KINOSITA (1957) have shown that the sex chromosomes bear the nucleolus organizer in the mouse. SACHS (1955), OHNO, KAPLAN and KINOSITA (1959) and GEYER-DUSZYNSKA (1963) assumed that the sex vesicle had a special material rich in RNA. SOLARI (1964) and SOLARI and TRES (1967a) showed that the two regions, the chromatinic and the nucleolar one, of the sex pair of the mouse could explain the previous optical observations without assuming the existence of a special material in the sex pair. The RNA content of the sex vesicle of mice is derived exclusively (at the histochemically defined level) from the attached nucleolus (SOLARI and TRES, 1967a). The sex pair does not have either granular regions or histochemically detectable RNA in species in which the nucleolar organizer is not associated with the sex chromosomes (although in some cells the nucleolus can lie beside the sex pair). The nucleolar part of the sex vesicle of the mouse begins developing at early and middle pachytene, coming to full size at late pachytene. During each of these stages the uptake of RNA precursors by the sex pair is null or very low (MONESI, 1965). Thus, SOLARI and TRES (1967a) assumed a special rRNA metabolism and transport in mouse spermatocytes, in accordance with the observations and hypothesis of DAS (1965) and DAS and ALFERT (1966).

The cores of the sex vesicle are permanent and constant structures of the sex pair. It is proposed here that each of the cores is the functional axis of each of the sex chromosomes during meiosis. This assumption is based on the following arguments. a) At very early pachytene (and probably zygotene) the long and the short cores are axial to the larger and smaller masses, respectively, of chromatin forming the sex pair. b) The length ratio of the cores (2.5) is nearly the same

FIGURE 3.—Reconstruction of the sex pair at early pachytene. Figure 3a is a photograph of the model of the complete sex pair. Figures 3b-3j are seriated micrographs of the sex pair. *c e,* common end of both cores; *l,* long core; *s,* short core.

as the length ratio of the X-Y pair at metaphase (OHNO and LYON, 1965). c) The cores are regularly associated at early pachytene in a synaptinemal complex in the common end. d) The cores behave in Searle's translocation as the hypothetical axis of the X chromosome would behave; that is, there is physical continuity between the long core (divided into two parts by the translocation) and a lateral component of a synaptinemal complex connected with the sex vesicle and autosomal in nature.

The relationship between the X and Y chromosomes in the mouse—and in mammals in general—has been debated during the past thirty years. KOLLER and DARLINGTON (1934) proposed that synapsis was restricted to the "homologous segment" of the sex chromosomes, the differential segments being unpaired. However, that idea was not sufficiently supported by cytological evidence, and it was vigorously contradicted by SACHS (1954, 1955) and by OHNO and WEILER (1962), although in a careful study of the sex pair of the mouse, OHNO, KAPLAN and KINOSITA (1959) suggested that the "end-to-end" relationship of the X-Y pair was the result of a chiasma between minute homologous regions. In our early study of the ultrastructure of the sex vesicle of the mouse (SOLARI, 1964) synaptinemal complexes were not observed, but it was specifically remarked that a *short, partial* synapsis could exist. That idea was rapidly supported by observations, as reported by SOLARI and TRES (1967b), of partial synapsis in the X-Y pair of the mouse. The present work confirms those observations, thus proving the essential part of the hypothesis of KOLLER and DARLINGTON (1934). The presence of a common end in every sex pair examined shows that synapsis is a regular feature of the sex pair. The formation of a synaptinemal complex at the common end shows that the behaviour of that part of the sex chromosomes is similar to that of homologous autosomes. The length of the synaptic region is, however, rather short; it amounts at early pachytene to less than 1/6 of the short core (assumed to be the Y axis), or, conversely, 1/16 of the long core. The sex chromosomes of the mouse seem to be mostly composed of differential segments; in the golden hamster (FORD and WOOLLAM, 1966) the pairing segment is longer. The possibility of crossing over is left open by the actuality of a synaptic region. Although evidence of partial sex linkage in mammals is scanty and controversial (GRÜNEBERG, 1952; HALDANE, 1936), it has been suggested (FERGUSON-SMITH, 1965) that crossing over in the human X-Y pair could help to explain some abnormal phenotypes.

The meaning of the synaptinemal complex as a morphological expression of pairing of homologues was suggested by MOSES (1956, 1958) and by MOSES and COLEMAN (1964). This subject has been also studied by ROTH (1966) and MOENS (1968). MEYER's (1961) observation of the lack of synaptinemal complexes in Drosophila males strongly supported its importance in synapsis and crossing-over. Although most of the data agree with Moses' hypothesis, there are some exceptional cases in which the presence of synaptinemal complexes may require a more elaborate explanation. For instance the observation by SOTELO and WETTSTEIN (1964) of synaptinemal complexes corresponding to the single X chromosome in Gryllus, the observation of multilayered bodies in Gryllus (SOTELO and

WETTSTEIN, 1965; GUENIN, 1965) and mosquito oöcytes (ROTH, 1966), and the observation of complexes in pollen mother cells of a haploid tomato by MENZEL and PRICE (1966) require a more complex explanation.

A similar exception to Moses' hypothesis may be found in what we call the *atypical* synaptinemal complex formed on the long core at middle pachytene. This transient structure, present on what is assumed to be the axis of the X chromosome, is difficult to explain. It is tentatively suggested that this atypical complex is formed between the two filaments composing the core at earlier stages and that each filament could represent axial structures of the X chromatids.

This work has been supported by a grant of the Consejo Nacional de Investigaciones Cientificas. I express my thanks to PROF. R. E. MANCINI for his support to this work and his interest in its achievement. I owe much help to DR. L. TRES during this work and I thank MISS M. LEMA for her able assistance.

LITERATURE CITED

DAS, N. K., 1965 Inactivation of the nucleolar apparatus during meiotic prophase in corn anthers. Exp. Cell Res. **40**: 360–364.

DAS, N. K., and M. ALFERT, 1966 Nucleolar RNA synthesis during mitotic and meiotic prophase. Nat. Cancer Inst. Monograph **23**: 337–351.

FERGUSON-SMITH, M. A., 1965 Karyotype-phenotype correlations in gonadal dysgenesis and their bearing on the pathogenesis of malformations. J. Med. Genetics **2**: 142–155.

FORD, E. H., and D. H. WOOLLAM, 1966 The fine structure of the sex vesicle and sex chromosome association in spermatocytes of mouse, golden hamster and field vole. J. Anat. **100**: 787–799.

GEYER-DUSZYNSKA, I., 1963 On the structure of the X-Y bivalent in *Mus musculus* L. Chromosoma **13**: 521–525.

GRUNEBERG, H., 1952 *The Genetics of the mouse*. Nijhoff, The Hague.

GUENIN, H. A., 1965 Observations sur la structure submicroscopique du complexe axial dans les chromosomes meiotiques chez *Gryllus campestris* L. et *G. bimaculatus* De Geer (Orthopt. Gryll.). J. Microscopie **4**: 749–758.

HALDANE, J. B. S., 1936 A search for incomplete sex linkage in man. Ann. Eugenics **7**: 317–326.

KOLLER, P. C., and C. D. DARLINGTON, 1934 The genetical and mechanical properties of sex chromosomes. I. *Rattus norvegicus.* J. Genet. **29**: 159–173.

MENZEL, M. Y., and J. M. PRICE, 1966 Fine structure of synapsed chromosomes in F1 *Lycopersicon esculentum-Solanum lycopersicoides* and its parents. Amer. J. Bot. **53**: 1079–1086.

MEYER, G. F., 1961 The fine structure of spermatocyte nuclei of *Drosophila melanogaster*. Proc. Europ. Reg. Conf. El. Microscopy **2**: 951–954.

MITTWOCH, U., 1967 *Sex chromosomes.* Academic Press, New York.

MOENS, P. B., 1968 The structure and function of the synaptinemal complex in *Lilium longiflorum* sporocytes. Chromosoma **23**: 418–451.

MONESI, V., 1965 Synthetic activities during spermatogenesis in the mouse. Exp. Cell Res. **39**: 197–224.

MOSES, M. J., 1956 Chromosomal structures in crayfish spermatocytes. J. Biophys. Biochem. Cytology **2**: 215–218. —— 1958 The relation between the axial complex of meiotic prophase chromosomes and chromosome pairing in a salamander (*Plethodon cinereus*). J. Biophys. Biochem. Cytology **4**: 633–638.

MOSES, M. J., and J. R. COLEMAN, 1964 Structural patterns and the functional organization of chromosomes. *The Role of Chromosomes in Development*, Edited by M. LOCKE. Academic Press, New York, pp. 11–49.

OAKBERG, E. F., 1956 A description of spermiogenesis in the mouse and its use in analysis of the cycle of the seminiferous epithelium and germ cell renewal Amer. J. Anat. **99**: 391–413.

OHNO, S., W. KAPLAN, and R. KINOSITA, 1957 Heterochromatic regions and nucleolus organizers in chromosomes of the mouse, *Mus musculus.* Exp. Cell Res. **13**: 358–364. —— 1959 On the end-to-end association of the X and Y chromosomes of *Mus musculus.* Exp. Cell Res. **18**: 282–290.

OHNO, S., and C. WEILER, 1962 Relationship between large Y chromosome and side-by-side pairing of the X-Y bivalent observed in the chinese hamster, *Cricetus griseus.* Chromosoma **13**: 106–110.

OHNO, S., and M. F. LYON, 1965 Cytological study of Searle's X-autosome translocation in *Mus musculus.* Chromosoma **16**: 90–100.

READER, C., and A. J. SOLARI, 1968 The histology and cytology of the seminiferous epithelium of the mouse with Searle's X-autosome translocation (Submitted for publication).

ROTH, T. F., 1966 Changes in the synaptinemal complex during meiotic prophase in mosquito oöcytes. Protoplasma **61**: 346–386.

SACHS, L., 1954 Sex linkage and the sex chromosomes in man. Ann. Eugen. **18**: 255–261. —— 1955 The possibilities of crossing-over between the sex chromosomes of the house mouse. Genetica **27**: 309–322.

SEARLE, H. G., 1962 Is sex-linked *tabby* really recessive in the mouse? Heredity **17**: 297.

SOLARI, A. J., 1964 The morphology and the ultrastructure of the sex vesicle in the mouse. Exp. Cell Res. **36**: 160–168. —— 1969 The evolution of the ultrastructure of the sex chromosomes (sex vesicle) during meiotic prophase in mouse spermatocytes. J. Ultrastructure Res. (In press).

SOLARI, A. J., and L. TRES, 1967a The localization of nucleic acids and the argentaffin substance in the sex vesicle of mouse spermatocytes. Exp. Cell Res. **47**: 86–96. —— 1967b The ultrastructure of the human sex vesicle. Chromosoma **22**: 16–31.

SOTELO, J. R., and R. WETTSTEIN, 1964 Electron microscope study on meiosis. The sex chromosome in spermatocytes, spermatids and oöcytes of *Gryllus argentinus.* Chromosoma **15**: 389–415. —— 1965 Fine structure of meiotic chromosomes. Nat. Cancer Inst. Monograph **18**: 133–152.

TRES, L., and A. J. SOLARI, 1968 The ultrastructure of the nuclei and the behaviour of the sex chromosomes in human spermatogonia. Z. Zellforsch. mikroskop. Anat. **91**: 75–89.

Meiosis in the Djungarian Hamster
I. General Pattern of Male Meiosis

HELEN E. POGOSIANZ

I. Introduction

The Djungarian or striped hairy-footed hamster (*Phodopus sungorus* Pall.) is a small rodent which can be easily bred under laboratory conditions (Meyer, 1967; Pogosianz and Sokova, 1967). This animal is a suitable tool for karyological studies. Its diploid chromosome number was estimated as 28 (Matthey, 1957) and most of its large and medium-size chromosomes (including the X and the Y) can be easily identified (Pogosianz and Brujako, 1967; Vorontsov, Radzabli, and Liapunova, 1967; Pogosianz, Sokova, and Yanovich, 1970).

The first meiotic observations of *Ph. sungorus* (Pogosianz and Brujako, 1969) have shown that meiosis in this species deserved a more detailed study. This paper presents a general description of male meiosis in *Ph. sungorus* with special attention to the assumed diffuse stage and to the sex-bivalent behaviour.

II. Material and Methods

Ten adult (2—3 months old) males of a middle-asiatic subspecies (*Phodopus sungorus campbelli* Thomas) from the breeding colony of our laboratory were used for the investigation. Chromosome preparations were obtained by the air-drying method of Evans, Breckon, and Ford (1964) sometimes without hypotonic pretreatment. Slides were stained with azur-eosin. Unfortunately in most cases the number of cells at the stages from diplotene to metaphase II was low, which should probably be attributed to the loss of some of the dividing cells during slide processing.

III. Results

Short Characterization of the Main Meiotic Stages. The leptotene stage of meiotic prophase in *Ph. sungorus* resembles this stage in other mammals. The nucleus is comparatively small and contains thin chromosome threads. Nuclei typical for the zygotene stage have not been observed. Pachytene nuclei were numerous. At this stage, when synapsis is already completed, the chromosomes (bivalents) become perceptibly thicker and acquire a characteristic woolly appearance (Fig. 1). In some nuclei fine lateral projections can be seen rather clearly. This is in agreement with the current view that the lamp-brush type of structure is inherent not only in amphibian oocyte chromosomes but in prophase meiotic chromosomes of different species of animals as well (Callan, 1963; John and Lewis, 1965). Usually the arrangement of bivalents inside the pachytene nucleus is random. Sometimes, however, both ends of particular chromosomes seem to adjoin the nuclear membrane (Fig. 1b).

The characteristic feature of pachytene nuclei is the presence of a dark body generally interpreted as the sex-vesicle. As has been shown for several mammalian species it consists of the heteropycnotic and condensed sex-chromosomes (Ford and Woollam, 1966 and others).

In *Ph. sungorus* the sex-vesicle has two zones, a dark and a light one. A similar fact has been described for the Golden and European hamsters (Fredga and Santesson, 1964). The dark zone is stained deeper and the light one less intensively than the autosomes. In *Ph. sungorus* the light zone made up of coiled double thread or band. In some cells this thread extends beyond the dark zone and looks like a loop or a "tail" (Fig. 1). In contrast to autosomes this tail has a smooth appearance.

An interesting feature of male meiosis in *Ph. sungorus* is the possible existence of the diffuse stage between pachytene and diplotene. We suppose that this stage is represented by nuclei of larger size than pachytene nuclei and which contain no clearly defined chromosomes (Fig. 2a, b).

Thin parallel chromatin threads which occasionally cross are seen in such nuclei. The process of pairing seems to be finished and homologous chromosomes begin to separate. The presence of several double chromatin

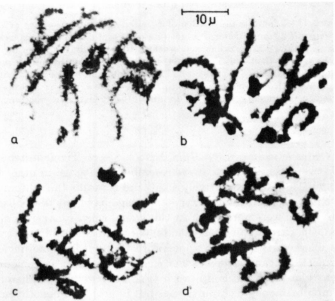

Fig. 1 a—d. Pachytene stage. a The sex-vesicle composed of two zones: a Dark and a light one; b—d the light zone in a form of a loop (b, c) or a "tail" (d)

knobs situated at identical sites of the threads also shows that these nuclei are at the post-pachytene stage.

The sex-bivalent is still present but it is not so distinctive here as at pachytene. This should probably be explained by the beginning of its decondensation and unfolding. In some nuclei the chromosome structures were more obvious (Fig. 2c). Such nuclei apparently are early diplotene stages.

The diffuse stage in male meiosis is probably of short duration because the number of the nuclei assigned to this stage was much lower than the number of pachytene nuclei on the same slides.

It should be noted that the large size of diffuse nuclei cannot be attributed to the action of hypotony because in material prepared without such pretreatment the nuclei of the same type and size were present.

At diplotene or early diakinesis (Fig. 2d) all 14 bivalents can be distinguished including the sex-bivalent with end-to-end paired X and Y chromosomes. There is no definite position or orientation of the sex-bivalent inside the cell. Bivalents are rather elongate, have woolly or chromomeric appearance (expressed, however, not so clearly as at pachytene), and a double structure.

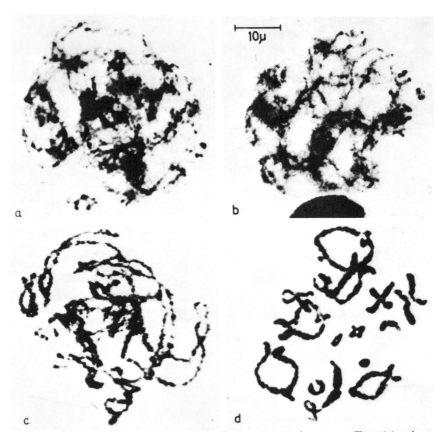

Fig. 2. a, b The assumed diffuse stage. Description in the text. c Transition from diffuse stage to diplotene. d Diplotene or early diakinesis

Large autosomes have usually 2—3, and medium-sized ones 1—2 chiasmata. The configuration of the bivalents varied from cell to cell, indicating non-localization of chiasmata. At diakinesis bivalents become especially clear-cut (Fig. 3a—c). Only the non-pairing arm of the X chromosome remains stretched (see below).

At metaphase I (Fig. 3d) the bivalents continue to contract and their double structure becomes less clearly visible than at diakinesis.

The individual identification of autosome bivalents, according to the classification suggested for somatic chromosomes of this species (Pogosianz and Brujako, 1967), is not reliable because after azur-eosin staining their centromeres are not readily revealed. However, one can distinguish

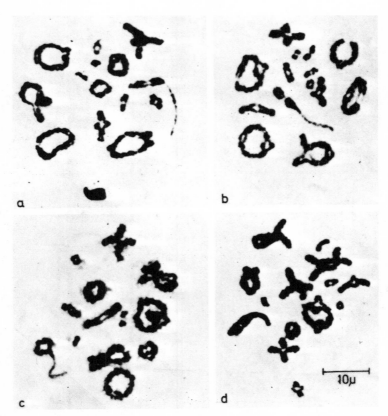

Fig. 3. a—c Diakinesis. d Metaphase I. Note the stretching of the non-pairing arm of X chromosome

without difficulty three groups of bivalents—large, medium—size and small ones, corresponding to the three groups of autosome pairs in somatic cell (Fig. 4).

At metaphase II (Fig. 5) chromosomes (dyads) have the appearance typical for this stage. As result of chromatid repulsion, they assume an X-like or V-like shape. At this stage of meiosis in some cells almost all chromosomes discernable in somatic cells can be identified. The spermatocytes II with X and with Y can be distinguished rather easily, the Y looks like a rod-shaped or V-shaped element; the X is cross-shaped and differs from the similar metacentric third chromosome by its smaller size. No heteropycnosis of the sex-chromosomes has been observed at this stage.

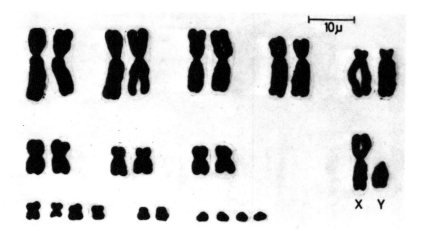

Fig. 4. Somatic chromosomes from a bone-marrow cell of a male *Ph. sungorus*

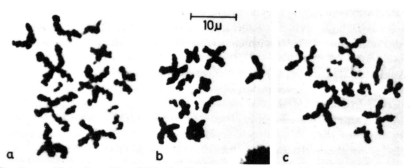

Fig. 5a—c. Metaphase II. a With the X chromosome; b, c with the Y chromosome

The Sex-Bivalent. The peculiar feature of the sex-bivalent in *Ph. sungorus* is the considerable delay of condensation of the non-pairing arm of the X chromosome during meiotic prophase. This allocyclic behaviour of the X is most clearly expressed at diplotene and diakinesis. The non-pairing (differential) arm of the X looks like a thin double thread (Figs. 2d; 3; 6). Usually at diplotene this arm is 4—5 times and at diakinesis 3—4 times longer than the pairing arm. In somatic cells the X chromosome of *Ph. sungorus campbelli* is metacentric (Fig. 4).

At metaphase I (Fig. 3d) the difference in length of the two arms of the X still exists and as a rule the X does not become metacentric. The degree of condensation of the differential arm at metaphase I is not

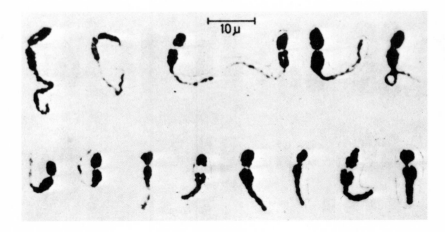

Fig. 6. Cut-out sex-bivalents. Progressive contraction of stretched differential arm of the X in cells advanced from diplotene to metaphase I

always correlated with the degree of condensation of other chromosomes in the same cells. In cells with comparatively short bivalents the X sometimes had a rather long "tail" (Fig. 3). Unlike the observation of Sasaki and Makino (1965) of sex-chromosome behaviour at metaphase I in man, precocious segregation of the sex-chromosomes has rarely been observed in *Ph. sungorus*.

If the allocycly of the X is clearly expressed already at diplotene one may suppose that this phenomenon has occurred at an earlier stage of meiotic prophase. Indeed, as mentioned above the light zone of sex vesicle at pachytene consists of a coiled pale thread, sometimes protruding from the vesicle as a loop or "tail". Since the dark zone includes pairing segments of the sex-chromosomes, the light thread should represent a non-condensed differential arm of the X. Thus, already at pachytene, this arm of the X differs in its appearance and probably in functional activity from other parts of the sex-bivalent.

In *Ph. sungorus* as in most of the mammalian species the first meiotic division is reductional for the sex-chromosomes. X and Y pair by their telomeric ends. Only in one out of about 300 spermatocytes I a lateral type of sex-chromosomes pairing was observed (Fig. 3c). It is not excluded that this lateral chiasma occurred accidentally or represents an artefact.

In conventional cytological preparations it is not possible to reveal the type of sex-chromosome pairing at pachytene, inside the sex vesicle. In Golden hamster where at diakinesis the X and Y also pair end-to-end (Fredga and Santesson, 1964) structures typical for the synaptonemal complex were observed in the sex vesicle with the aid of electron microscopy (Ford and Woollam, 1966). These structures were revealed in both dark and light zones indicating that the lateral pairing must have taken place along the whole length of the sex-chromosomes. In *Ph. sungorus* one of the arms of X clearly does not participate in pairing at pachytene because the whole Y is equal to only one arm of the X and the differential arm is sometimes free and protudes from the vesicle. The question of the type of sex-chromosome pairing (lateral or terminal) inside the dark zone of the vesicle in *Ph. sungorus* is open for further investigations.

Discussion

Although meiosis in *Ph. sungorus* in the sequence of the main stages and in the prereduction of the sex-chromosomes seems to be similar to that of most mammalian species so far studied, two of its features deserve special attention. First, the supposed existence of the diffuse stage between pachytene and diplotene. As far back as 1925, Wilson noted that the diffuse stage is inherent in meiosis both in females and males, but in oocytes the degree of decondensation of the chromosomes is deeper than in spermatocytes.

In the oogenesis of some mammals (mouse, rat) the diffuse stage takes an extreme form of so called dichtyatene (Beaumont, 1968; Beaumont and Mandl, 1962; John and Lewis, 1965).

Recently the diffuse stage has been described for male meiosis of several organisms (Barry, 1969, Moens, 1964; Sen, 1969). However, in interpretation of the relevant cytological picture and in locating this stage in meiotic prophase the opinions diverge. Some authors believe that the diffuse stage precedes pachytene (Peakock, 1968), others that it occurs between pachytene and diplotene (Barry, 1969; Moens, 1964, 1968). A third group of authors think that this stage follows diplotene (Beaumont, 1968, Beaumont and Mandl, 1962). Moens (1964) points out that the stage which has been described for some plants as zygotene is in fact a diffuse stage which take place after pachytene.

This contradiction is probably due both to differences in the course of meiosis in different organisms and to the insufficient knowledge of the details of this process.

The diffuse stage in male hamsters has not been described. Moreover, in the recent symposium on effects of radiation on meiotic systems (1968) there is no reference to the existence of such a stage in mammalian spermatogenesis.

In male meiosis of *Ph. sungorus* no final conclusions on the time and place of this stage in meiotic prophase can be drawn from the study of air-dried preparations of disintegrated tissue. A more definite answer to these questions should be obtained from sections. Indirectly, on the basis of the large size of the nucleus, the presence of remanent sex-vesicle and double, occasionally crossed, thin chromatin threads with double chromatin knobs, one may assume that this stage follows pachytene and precedes diplotene.

The physiological significance of the diffuse stage (the period of a high metabolic activity or chromosome decondensation important for crossing-over, etc.) is still unknown (see Barry, 1969). Further studies should clarify the time of occurence, importance and degree of universality of a diffuse stage in male meiosis of different mammalian species.

The second interesting feature of male meiosis in *Ph. sungorus* concerns the sex-bivalent. Typical of this species is the allocycly of the differential arm of the X, which is manifested in the delayed condensation of this arm during the meiotic prophase. The difference in the degree of contraction of two arms of the X is decreased (but usually does not disappear completely) only at metaphase I.

As has been shown with the aid of H^3-thymidine autoradiography, the whole Y and one arm of the X chromosome are late replicating in somatic cells of *Ph. sungorus* (Pogosianz, Sokova, and Yanovich, 1970). The question arises, which of the two arms of the X — the pairing or the differential one — is heterochromatic (late replicating) and functionally inactive in somatic cells ? If negative heteropycnosis is not a synonym of the heterochromatin (Lima-de-Faria and Jaworska, 1968) then it may be assumed that the homologous (pairing) arm is late replicating and the differential arm (which is negatively heteropycnotic at pachytene) is early replicating, i. e. functionally active in somatic cells.

This problem, however, cannot be solved without a special autoradiographic investigation of meiotic chromosomes in this species because in several mammalian species a change of the replication pattern has been found in spermatogonia as compared with somatic cells (Utakoji and Hsu, 1965; Mukherjee and Ghosal, 1969). It is of interest that the phenomenon of stretching of one arm of the X in meiotic prophase has been observed also in the Golden hamster (Koller, 1938; Schaeffer, 1955; Emmons and Husted, 1962; Fredga and Santesson, 1964). Similar phenomena have been revealed for both X and Y chromosomes of the European hamster (Fredga and Santesson, 1964). The comparative study of male meiosis in three hamster species has demonstrated specific differences in their sex-bivalents. A schematic representation of the sex-bivalents in 4 hamster species (Fig. 7) shows that all four differ from each other. In *C. griseus* and *C. cricetus* the type of sex-chromosome pairing is lateral, and in *M. auratus* and *Ph. sungorus* it is terminal. In *C. griseus* there is no delay in condensation of the differential arm of the

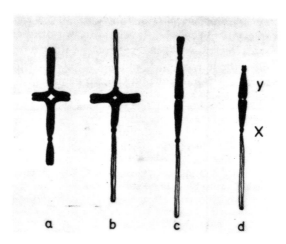

Fig. 7 a—d. Schematic representation of sex-bivalents in 4 hamster species. a *C. griseus*; b *C. cricetus*; c *M. auratus*; d *Ph. sungorus*. a According to Utakoji (1966); b, c according to Fredga and Santesson (1964); c According to the present work. Recently Fraccaro, Gustavsson, Hulten, Lindsten and Tiepolo (1969) confirmed the data of Fredga and Santesson (1964) that in *C. griseus* the X pairs by its short arm. This difference from the data of Utakoji (1966) is of no importance for the given comparison

X, whereas in *C. cricetus* both X and Y chromosomes display this phenomenon. *Ph. sungorus* is nearer *M. auratus* in this respect. The difference in form of the sex-bivalent in these two species is due to the different types of the Y: it is telocentric in *sungorus* and biarmed in *auratus*. The size of this bivalent in the two species is probably also somewhat different because in *auratus* the X is the biggest chromosome in the whole set and in *sungorus* it is only the 6th in length.

On the basis of recent data on the relationship between the heterochromatinization, the degree of condensation, time of replication and functional activity of the chromosome (Lima-de-Faria and Jaworska, 1968; Zakharov, 1968) one may assume that delayed condensation of the nonpairing arm of the X chromosome reflects its active functional state in meiotic prophase. This activity may be connected with the control of synthesis of products important for a normal course of meiosis.

From this point of view it remains unclear, however, why the phenomenon of sex-chromosome allocycly (or partial allocycly) has not been demonstrated in all mammalian species. In mouse, rat, steppe-lemming and in man it seems not to exist. In some other mammals, it does exist as can be seen from illustrations given by Matthey (1961) and Fredga (1968). The possibility is not excluded that in those species where this allocycly has not been found the functional activity of sex chromosomes in meiotic prophase is either quite different or it differs in time, i. e. it ceases earlier than in those species where the allocycly was revealed. Comparative electron microscopic, radioautographic, cytochemical and other studies may shed some light on this problem.

Acknowledgement. The author is grateful to E. T. Brujako for her help in breeding of hamsters and for valuable technical assistance.

References

Barry, E. G.: The diffuse diplotene stage of meiotic prophase in *Neurospora*. Chromosoma (Berl.) **26**, 119—129 (1969).
Beaumont, H. A.: Radiosensitivity of primordial oocytes in the rat and monkey. In: Effects of radiation on meiotic systems, p. 71—79 Vienna: IAEA 1968.
— Mandl, A. M.: A quantitative and cytological study of oogonia and oocytes in the foetal and neonatal rat. Proc. roy. Soc. Lond. B **155**, 552—579 (1962).
Callan, H. C.: The nature of lampbruch chromosomes. Int. Rev. Cytol. **15**, 1—34 (1963).
Effects of radiation on meiotic systems. Vienna: IAEA 1968.
Emmons, L. R., Husted, L.: The sex bivalent of the Golden hamster. J. Hered. **53**, 227—232 (1962).
Evans, E. P., Breckon, G., Ford, C. E.: An air drying method for meiotic preparations from mammalian testes. Cytogenetics **3**, 289—294 (1964).
Ford, E. H. R., Woollam, D. H. M.: The fine structure of the sex vesicle and sex-chromosome association in spermatocytes of mouse, golden hamster and field vole. J. Anat. (Lond.) **100**, 787—799 (1966).
Fraccaro, M., Gustavsson, I., Hultén, M., Lindsten, J., Tiepolo, L.: Late-replicating Y chromosome in spermatogonia of the Chinese hamster (*Cricetulus griseus*). Cytogenetics **8**, 263—271 (1969).
Fredga, K.: Idiogram and trisomy of the water vole (*Arvicola terrestris* L.), a favourable animal for cytogenetic research. Chromosoma (Berl.) **25**, 75—89 (1968).
— Santesson, K. B.: Male meiosis in Syrian, Chinese, and European hamsters. Hereditas (Lund) **52**, 36—48 (1964).
John, B., Lewis, K. R.: The meiotic system. Protoplasmatologia VI/F/1. Wien-New York: Springer 1965.
Koller, P. C.: The genetical and mechanical properties of the sex-chromosomes. IV. The golden hamster. J. Genet. **36**, 177—195 (1938).
Lima-de-Faria, A., Jaworska, H.: Late DNA synthesis in heterochromatin. Nature (Lond.) **127**, 138—142 (1968).
Matthey, R.: Analyse cytotaxonomique de huit espèces de *Muridae, Murinae, Cricetinae, Microtinae* paléaectiques et nord-américains. Arch. Klaus-Stift. Vererb.-Forsch. **32**, 385—404 (1957).
— Cytologie comparée des *Cricetinae* paléarctiques et américains. Rev. suisse Zool. **68**, 41—61 (1961).
— La formule chromosomique chez sept espèces et sous-espèces de *Murinae* africains. Mammalia (Paris) **27**, 157—176 (1963).
Meyer, M. N.: Peculiarities of the reproduction and development of *Phodopus sungorus* Pallas of different geographical populations. Zool. J. (USSR) **46**, 604—613 (1967).

Moens, P. B.: A new interpretation of meiotic prophase in *Lycopersicon esculentum* (tomato). Chromosoma (Berl.) 15, 231—242 (1964).
— The structure and function of the synaptinemal complex in *Lilium longiflorum* sporocytes. Chromosoma (Berl.) 23, 418—451 (1968).
Mukherjee, B. B., Ghosal, S. K.: Replicative differentiation of mammalian sex-chromosomes during spermatogenesis. Exp. Cell Res. 54, 101—106 (1969).
Peacock, W. J.: Replication, recombination and chiasmata in *Goniaea australasiae* (*Orthoptera: Acrididae*). (1968) (Cit. from Barry, 1969).
Pogosianz, H. E., Brujako, T.: Somatic chromosomes of Djungarian hamster (*Phodopus sungorus*). Genetics (USSR) 3, 2—20 (1967).
— — Polyploid cells in mamalian meiosis. Genetics (USSR), 5, 176—178 (1969).
— Sokova, O. I.: Maintaining and breeding of the Djungarian hamster under laboratory conditions. Z. Versuchstierk. 9, 292—297 (1967).
— — Yanovich, L. I.: The normal karyotype of the Djungarian hamster. Cytologia (USSR) (in press) 1970.
Sasaki, M., Makino, M.: The meiotic chromosomes of man. Chromosoma (Berl.) 16, 637—651 (1965).
Sen, S. K.: Chromatin-organization during and after synapsis in cultured microsporocytes of *Lilium* in presence of mitomycin C and cyclohexinide. Exp. Cell Res. 55, 123—127 (1969).
Sheaffer, C. I.: The X-bivalent of the golden hamster. Virginia J. Sci. 6, 46—52 (1955).
Utakoji, T.: On the homology between the X and the Y chromosomes of the Chinese hamster. Chromosoma (Berl.) 18, 449 (1966).
— Hsu, T. S.: DNA replication patterns in somatic and germline cells of the male Chinese hamster. Cytogenetics 4, 295—305 (1965).
Vorontsov, N. N., Radzabli, S. L., Liapunova, K. L.: Caryological differentiation of allopatric forms of hamsters species *Phodopus sungorus* and heteromorphism of sex-chromosomes in females. Dokl. Acad. Sci. USSR (Ser. B) 172, 703—705 (1967).
White, M. J. D.: The chromosomes, 5th ed. London: Methuen 1961.
Wilson, E. B.: The cell in development and heredity, 3rd ed. New York: Macmillan 1925.
Zakharov, A. F.: Heterochromatin and genetic inactivation in mammalian cells. Usp. sovrem. Biol. (USSR) 65, 83—106 (1968).

The Behaviour of Chromosomal Axes during Diplotene in Mouse Spermatocytes

ALBERTO J. SOLARI

Introduction

The synaptonemal complex is a tripartite, planar structure characteristic of the synaptic stages of meiotic chromosomes (Moses, 1958, 1968, 1969). Although the participation of the synaptonemal complex in chiasma formation has been suggested by several authors (Woollam and Ford, 1964; Meyer, 1964), a systematic study of its behaviour during diplotene is lacking in the literature. The relevance of the synaptonemal complex in chiasma formation has been underestimated because of the erratic behaviour of its lateral elements during diplotene in some species (Roth, 1966; Moens, 1968). However, there is strong evidence that at least in several mammals the elements of the synaptonemal complex persist during diplotene and have normal features (Franchi and Mandl, 1962; Baker and Franchi, 1967).

The aim of this paper is to study the changes that the elements of the synaptonemal complex show during diplotene in mouse spermatocytes. As the behaviour of the axes of the sex chromosomes has been previously described (Solari, 1969a, 1969b, 1970), this work will deal with the behaviour of the autosomal axial structures. Previous observations on the sex chromosomes (Solari, 1970) suggested that chromosomal axes do not cross-over in chiasmata. This assumption is proved by the present work.

Material and Methods

Albino mice from two stocks were used. Squashed spermatocytes at diplotene showed, as usual, that every bivalent had at least one chiasma.

Electron Microscopy. Pieces of testes were fixed in 2.5% glutaraldehyde in 0.1 M phosphate buffer, pH 6.9, and postfixed in Caulfield's fixative. Embedding was made in maraglas. Serial sections were cut and mounted on single hole grids (1×2 mm in diameter) according to Sjöstrand's (1967) method.

Two kinds of serial sections were used in this work: a) thin (silver coloured), nominally 500—750 Å thick; and b) intermediate (nominally from 1,000 to 2,500 Å in thickness).

Three-dimensional models of the chromosomal axes were made as previously described (Solari, 1970) in plastic sheets.

The labelling of the meiotic stage (diplotene) was based on the method previously described (Solari, 1969a), that is, on the determination of the stage of the spermatogenic cycle in pairs of adjacent thick (for examination with the light microscope) and thin sections.

Results

1. The Chromosomal Axes during Diplotene

Synaptonemal complexes are very seldom observed during diplotene; most axial structures of chromosomes are single (Fig. 1). Single axes are readily seen in sections of intermediate thickness, and they can be several microns long. Thus, diplotene nuclei are strikingly different from leptotene and zygotene nuclei with respect to their axes (in mouse spermatocytes single axes are difficult to observe and rather short during these latter stages, Solari, 1969a). The study of the whole path of the single axes with serial sections (see below) has shown that each axis has two structurally different ends: the basal knob end and the simple end. Thus, ends of both kinds, the dense and the non-dense ones, can be seen in diplotene nuclei (Figs. 1 and 2).

Each single axis appears as a well-defined, slender rod about 300 Å wide, that is, similar to the lateral elements of the synaptonemal complex (Fig. 2). Chromatin fibrils, 100—200 Å wide, are found on both sides of a longitudinal section of an axis. However, there is some pattern of alternate thickenings of the chromatin on both sides which gives the result that at some points the axis seems to be peripheral to the chromatin mass.

2. The Path of the Single Axes

The single axes begin and end on the nuclear membrane. Their path is generally smoothly curved, except in the convergence regions (see below) where the degree of curvature may increase sharply. The axes are continuous all over their path, as seen in serial sections (Fig. 3a, b and c). Cross-sections of the axes are difficult to distinguish from some dense granular component which is mixed with the chromatin fibrils.

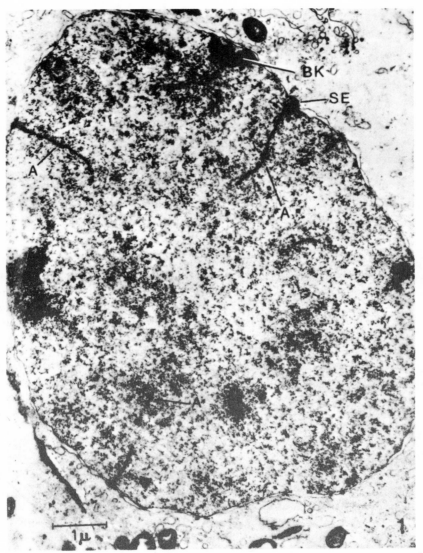

Fig. 1. Electron micrograph of a diplotene nucleus showing the single axes (*A*) ending on basal knobs (*BK*) or on simple ends (*SE*)

No cross-over of two axes was observed, neither showed the axes branches, with the exception of their ends. Where the axes come into the basal knobs they branch into 3 or 4 rods which frequently rejoin be-

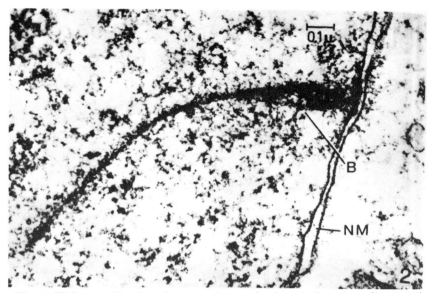

Fig. 2. High magnification electron micrograph of a single axis ending on the nuclear membrane (*NM*) by a simple end, the only places where branches (*B*) of the axis are seen

fore touching the nuclear membrane (Fig. 10). Although the branches are especially prominent inside the basal knobs, similar branchings occur at the simple ends (Fig. 2), that is, the ends surrounded by dispersed chromatin.

3. Convergence Regions of the Axes

Single axes have been observed to approach each other without crossing each other. These approachments or *convergence regions* can be located interstitially or at the end of the axes (Figs. 4—8).

Two kinds of structures have been observed at the convergence regions 1) a piece of a synaptonemal complex; and 2) a bridge of chromatin fibrils. When a piece of synaptonemal complex is formed in the convergence region (Figs. 4—6), each single axis continues itself with the lateral element of that piece and then again comes free (if it is an interstitial convergence region) as a single axis. This kind of convergence region is not frequent during late diplotene; it is observed during early diplotene, when such regions are also longer.

The most common structure at the convergence region is a chromatin bridge (Figs. 3, 7, 8 and 10). The axes come near each other and at

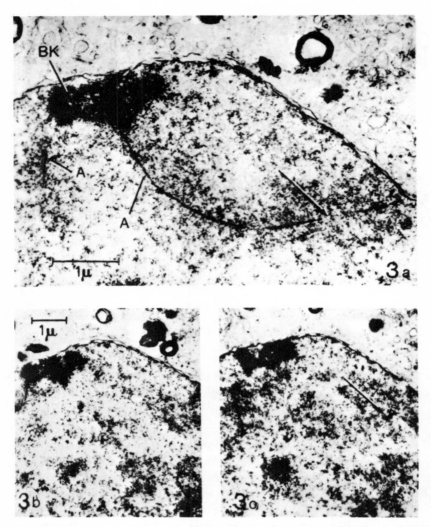

Fig. 3a—c. Serial electron micrographs showing the whole path of a single axis (*A* to the right). *BK* basal knob. Figure 3a shows almost completely the axis; the arrow marks a segment which is found in micrograph 3c. The left axis (*A* to the left) also ends in the same basal knob (Fig. 3b)

that point the space between them is filled with chromatin fibrils identical to those that accompany the axes at their sides (Figs. 7 and 8). The space filled by this chromatin bridge can become enlarged up to 0.6 to 0.8 μ.

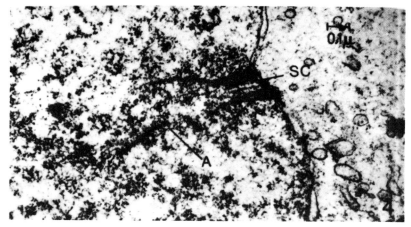

Fig. 4. A convergence region of the single axes at a simple end, where they form a synaptonemal complex (*SC*). The central region of the complex shows fibrils. *A* singles axis

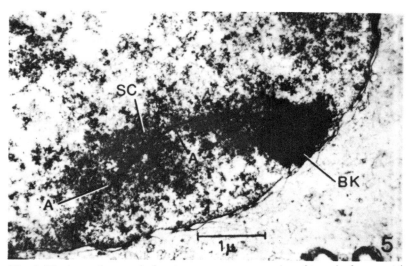

Fig. 5. Electron micrograph showing a convergence region of two singles axes (*A* and *A'*) where a piece of a synaptonemal complex is formed (*SC*). *BK* basal knob

Transitional states from a synaptonemal complex to a chromatin bridge have been observed. First, the central region of the synaptonemal complex becomes more dense, seemingly because of the appearance of chromatin fibrils in it (Fig. 9). Second is the disappearance of the central

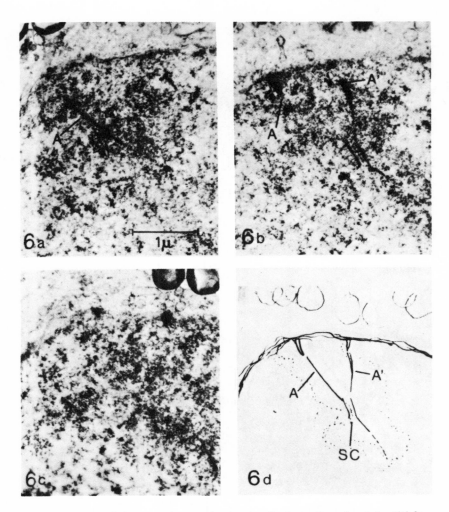

Fig. 6a—d. Serial electron micrographs showing the formation of an interstitial convergence region of two single axes (*A* and *A'*) where a piece of a synaptonemal complex is found. Fig. 6d is a photograph of the reconstructed model. *SC* piece of a synaptonemal complex

element and the central region, which becomes filled with chromatin fibrils (Fig. 10). Finally, the lateral elements separate from each other but remaining attached by the chromatin bridge. It has not been possible to ascribe the formation of the chromatin bridge to any particular part of the chromatin surrounding the single axes.

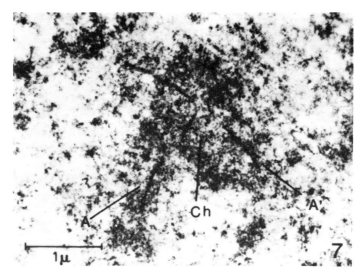

Fig. 7. Electron micrograph of a convergence region of two axes (A and A') where only a chromatin bridge (Ch) is found

4. Irregularities of the Axes

Very few cases of irregular axes were observed. The axes were not detached from chromatin masses and they were continuous. A single irregular behaviour, consisting in the blind ending of an axis on a dense body about 0.3 μ wide was observed.

The fate of the axes was not determined. Diakinesis is a very short stage difficult to observe. Metaphase chromosomes do not evidence any axes, with the exception previously described (Solari, 1969a) of one chromosome.

Discussion

1. The Behaviour of the Synaptonemal Complex at the End of Pachytene

The fate of the synaptonemal complex after pachytene is not clear, as conflicting data have been reported in different species (Moses, 1968). In a comprehensive review Moses (1968) concludes that at least in certain vertebrates the synaptonemal complex disappears by separation into two axial components, and thus no synaptonemal complexes would exist during diplotene.

In some species, as the salamander (Moses, 1958) and the grasshopper (Sotelo and Trujillo-Cenoz, 1960) no traces of single axes have been described during diplotene. However, early condensation of chromosomes during diplotene has been reported in the salamander (Moses, 1958) and this process could mask the single axes.

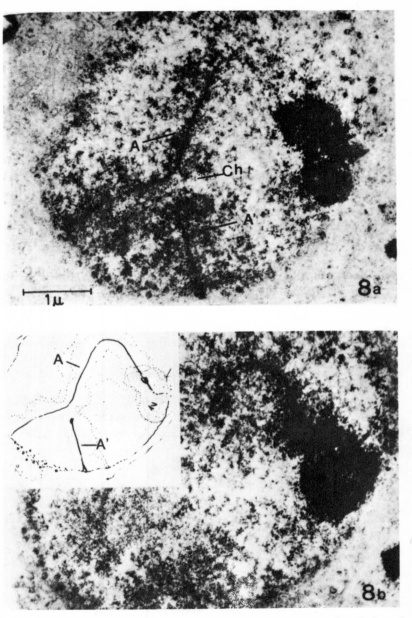

Fig. 8a and b. Electron micrographs of sequential sections showing the whole path of one axis (A) and part of its homologous axis (A'). The interstitial convergence region shows a chromatin bridge (Ch). Inset: reconstructed model

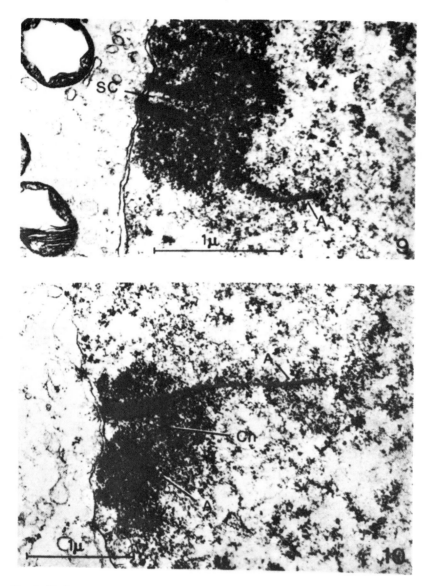

Fig. 9. Early stage of a convergence region in a basal knob. *A* axis; *SC* piece of a synaptonemal complex with the central element fading

Fig. 10. Later stage of a convergence region in a basal knob. *A*, *A'* axes; *Ch* chromatin bridge

In other species, as the lily (Moens, 1968) and the mosquito (Roth, 1966), single axes were described but their attachment to chromatin masses has been reported to be irregular.

However, in rat (Franchi and Mandl, 1962), mouse (Tsuda, 1965) and human (Baker and Franchi, 1967) oocytes clear evidence of the presence of single axes in well identified diplotene chromosomes has been reported. The present observations agree with the latter about the presence of single axes during diplotene. However, Baker and Franchi (1967) suggested that each single axis is formed by two filaments, and that the pattern of the chromatin fibrils around the axes is of the lampbrush type. Our results in mouse spermatocytes differ from these suggestions. Doubleness of the single axes is not evident at any place except when branching at the ends (see p. 219/220). The lampbrush pattern of chromatin fibrils was not evident in mouse spermatocytes, although it cannot be discarded. The relationship between the single axes seen with the electron microscope and the axes of lampbrush-type chromosomes (Callan, 1963) is not fully clear. As lampbrush chromosomes are characteristic of late diplotene, and during the last part of diplotene or diakinesis the single axes must be masked or disappear in mouse spermatocytes, both structures may not be identical.

The synaptonemal complex has been proved to extend over the whole length of pachytene chromosomes (Wettstein and Sotelo, 1967). The present results show that the single axes have large similarities with the lateral elements of the synaptonemal complex. They are continuous all over each chromatin mass, and they are attached at both ends to the nuclear envelope, as the synaptonemal complexes during pachytene (Woollam, Ford and Millen, 1966). These facts, and the observation of the remains of synaptonemal complexes in the convergence regions (see below) show that synaptonemal complexes partially disappear by the separation of its lateral elements at the end of pachytene in mouse spermatocytes. Two differences have been observed between the lateral elements and the single axes: 1) the chromatin is more sparsely distributed around the single axes, and its pattern is different, because it surrounds the axes; and 2) the occasional doubleness of the lateral elements is not evident in the single axes.

The branching at the ends of the single axes is similar to the opening of the lateral elements under hypotonic shock (Woollam and Ford, 1964).

2. *The Structure of the Convergence Regions and the Participation of the Axes in Chiasmata*

The axes or cores of the XY pair in mouse spermatocytes are structurally and functionally similar to the single axes of autosomes

(Solari, 1969a). From a study of the former (Solari, 1970) it was suggested that axes probably do not cross each other in chiasmata. The present results show that such a crossing does not occur either in the autosomes.

When the remains of the synaptonemal complex have completely disappeared, the only joining between homologues is a bunch of fibrils which are indistinguishable from those placed at the sides of synaptonemal complexes, which have the histochemical properties of chromatin (Coleman and Moses, 1964). The sequence of events leading to the formation of a convergence region of the axes with a chromatin bridge, may be assumed from the present results. At some moment during the end of pachytene, the lateral elements begin to separate from each other, preceded by fading of the central element of the synaptonemal complex and the appearance of chromatin fibrils on the medial side of the elements. In a few cases it has been observed that the largest distance separating the axes was spanned between the simple ends of the pair. However, the origin of the repulsion between the homologous axes may not be localized at specific points but may be extended along the chromosomes.

At some points, homologous axes remain near each other, connected by a piece of a synaptonemal complex. This piece gradually transforms into a chromatin bridge, which remains as such during late diplotene.

From the present results it has been inferred that the *convergence regions* of the axes are chiasmata. This inference is based mainly on the fact that each bivalent must have at least one chiasma, and the convergence regions are the only ones along the whole path of the cores where they approach each other with their accompanying chromatin fibrils. Besides, the path of the cores and their chromatin is similar to the chiasmatic configurations of bivalents.

The question may be raised if the pieces of synaptonemal complexes represent genuine chiasmata or they are only patches of unseparated homologues before repulsion. The transitional stages described between the pieces of synaptonemal complexes and the chromatin bridges strongly argue that they represent chiasmata during its first phase. Furthermore, such pieces can be located at very small and highly curved convergence regions, where, unless a particular steric hindrance for separation exists, it would be improbable that they are not yet separated. However, several arguments support the idea that these pieces of synaptonemal complexes must not be identical to the original complex. In fact, in most cases the central region of the piece was denser than that of a pachytene complex. Besides, the presence of a synaptonemal complex is not sufficient to insure the presence of chiasmata when that complex is not already located at a convergence region. Thus, Gassner (1969) has observed complexes during prophase in the achiasmatic meiosis of *Bolbe nigra*, and Roth and Parchman (1967) have shown that the presence of complexes during pachytene in *Trillium* microsporocytes is not sufficient for the formation of chiasmata during diplotene.

The fact that the single axes do not cross each other, and the lack of evidence about their duplicity strongly suggest that they do not represent the axes of chromatids (see also Solari, 1970). The observations on lampbrush chromosomes (Callan, 1963) suggest that chiasmata are always formed at the axes of those chromosomes and never at the loops. These apparently contrasting data may be fitted into hypothetical models of chiasma formation that will be dealt with elsewhere.

Acknowledgements. I thank Dr. L. Tres, Professor R. Mancini and Miss M. Lema for help and support during this work. This work was supported by a grant of the Population Council and grant TW 00317 of the National Institutes of Health. The author is member of the Scientific Career, Consejo Nacional de Investigaciones.

References

Baker, T. G., Franchi, L. L.: The structure of the chromosomes in human primordial oocytes. Chromosoma (Berl.) **22**, 358—377 (1967).

Callan, H. G.: The nature of lampbrush chromosomes. Int. Rev. Cytol. **15**, 1—34 (1963).

Coleman, J. R., Moses, M. J.: DNA and the fine structure of synaptic chromosomes in the domestic rooster (*Gallus domesticus*). J. Cell Biol. **23**, 63—78 (1964).

Franchi, L. L., Mandl, A. M.: The ultrastructure of oogonia and oocytes in the foetal and neonatal rat. Proc. roy. Soc. Lond. B **157**, 99—114 (1962).

Gassner, G.: Synaptinemal complexes in the achiasmatic spermatogenesis of *Bolbe nigra* Giglio-Tos (*Mantoidea*). Chromosoma (Berl.) **26**, 22—34 (1969).

Meyer, G. F.: Possible correlation between submicroscopic structure of meiotic chromosomes and crossing-over. Third Europ. Conf. on Electron Microscopy B, 461—462 (1964).

Moens, P. B.: The structure and function of the synaptinemal complex in *Lilium longiflorum* sporocytes. Chromosoma (Berl.) **23**, 418—451 (1968).

Moses, M. J.: The relation between the axial complex of meiotic prophase chromosomes and chromosome pairing in a salamander (*Plethodon cinereus*). J. biophys. biochem. Cytol. **4**, 633—638 (1958).

— Synaptinemal complex. Ann. Rev. Genet. **2**, 363—412 (1968).

— Structure and function of the synaptonemal complex. In: Nuclear physiology and differentiation. Genetics, Suppl. **61**, 41—51 (1969).

Roth, T. F.: Changes in the synaptinemal complex during meiotic prophase in mosquito oocytes. Protoplasma (Wien) **61**, 346—386 (1966).

—, Parchman, L. G.: Diplotene achiasmatic chromosomes following normal synapsis at pachynema. Proc. 25th Meeting Electron Micr. Soc. (C. J. Arceneaux, ed.), p. 86—87. Baton Rouge, La.: Claitor's 1967.

Sjöstrand, F. J.: Electron microscopy of cells and tissues. I. Instrumentation and and techniques. New York: Academic Press 1967.

Solari, A. J.: The evolution of the ultrastructure of the sex chromosomes (sex vesicle) during meiotic prophase in mouse spermatocytes. J. Ultrastruct. Res. **27**, 289—305 (1969a).

Solari, A. J.: Changes in the sex chromosomes during meiotic prophase in mouse spermatocytes. In: Nuclear physiology and differentiation. Genetics, Suppl. 61, 113—120 (1969b).
— The spatial relationship of the X and Y chromosomes during meiotic prophase in mouse spermatocytes. Chromosoma (Berl.) 29, 217—236 (1970).
Sotelo, R. J., Trujillo-Cenoz, O.: Electron microscopic study on spermatogenesis. Chromosome morphogenesis at the onset of meiosis (cyte I) and nuclear structure of early and late spermatids. Z. Zellforsch. 51, 243—277 (1960).
Tsuda, H.: An electron microscope study on the oogenesis in the mouse, with special reference to the behaviours of oogonia and oocytes at meiotic prophase. Arch. histol. jap. 25, 533—555 (1965).
Wettstein, R., Sotelo, R. J.: Electron microscope serial reconstruction of the spermatocyte I nuclei at pachytene. J. de Microsc. 6, 557—576 (1967).
Woollam, D. H., Ford, D. H.: The fine structure of the mammalian chromosome in meiotic prophase with special reference to the synaptinemal complex. J. Anat. (Lond.) 98, 163—173 (1964).
— — Millen, J.: The attachment of pachytene chromosomes to the nuclear membrane in mammalian spermatocytes. Exp. Cell Res. 42, 657—661 (1966).

INTERCELLULAR BRIDGES AND SYNCHRONIZATION OF GERM CELL DIFFERENTIATION DURING OOGENESIS IN THE RABBIT

LUCIANO ZAMBONI and BERNARD GONDOS.

A certain degree of synchronization of germ cell differentiation is a general feature of mammalian oogenesis. In the rat (3, 10), mouse (4), and rabbit (15, 16), germ cell differentiation is highly synchronized in that there is a direct correspondence between fetal age and predominant stage of cellular activity. In the guinea pig (11), monkey (2), and man (1), the synchronized pattern is less apparent because of a considerable overlapping of mitosis, meiosis, and degeneration. Correspondence between age and stage of differentiation of a significant percentage of germ cells is found even in these species, however.

In the course of an electron microscopic study of ovarian development in the newborn rabbit, we have observed that the synchronization of the oogenetic process finds expression not only in the close relationship between fetal age and stage of cellular differentiation, but also in the maturation of the germ cells in groups and in the synchronous differentiation of all the cells in each group. Such synchronization appears to be related

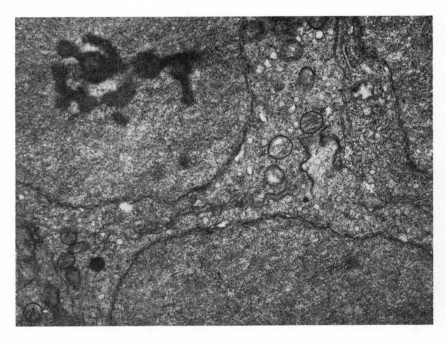

FIGURE 1 Two oogonia connected by an intercellular bridge. The bridge is limited by a plasma membrane of increased electron opacity. × 22,500.

to the presence of intercellular bridges connecting adjacent germ cells and resulting in a syncytial organization of the germ cells in the cords.

MATERIAL AND METHODS

This study was performed on the ovaries of 22 newborn New Zealand white rabbits, 1–15 days of age. The tissue fragments were fixed in 1% OsO$_4$ with salts added (17) and were embedded in Epon 812 (13). Sections were stained with lead hydroxide (12) and examined with Hitachi HU-11A and HU-11C electron microscopes.

RESULTS AND DISCUSSION

From days 1–3 postpartum the rabbit ovary is characterized by large numbers of mitotic oogonia which are consistently found in groups of two to six elements, all at the same stage of mitosis. In no instance could we find a dividing cell which was not in phase with the others in its group. At days 4 and 5 postpartum, when meiosis follows mitosis as the predominant activity of the germ cells, the meiotic cells are found in clusters consisting of oocytes which are all at the same stage of meiotic prophase. The waves of degeneration, which last until about day 15 and destroy great numbers of oocytes, also occur in a zonal pattern and involve groups of oocytes showing identical, regressive changes.

The structural basis for this synchronization appears to be a network of cytoplasmic bridges connecting adjacent germ cells to form syncytial groups. These intercellular bridges (Figs. 1–8), short and cylindrical in shape, are limited by a plasma membrane which is directly continuous with the plasma membranes of the connected cells but is distinguished from them by a greater electron opacity and an increased thickness.

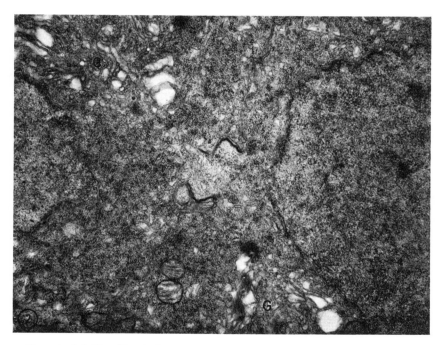

FIGURE 2 As in Fig. 1. Note the close morphological similarity of the two connected oogonia. A mirror-image configuration is indicated by the symmetrically opposite locations of the Golgi complexes (*G*) in the cells. × 19,500.

Identification of the bridges is thus possible even when the bodies of the joined cells do not appear in the section (Fig. 5). The cytoplasm of the bridges is identical with that of the connected cells (Fig. 4), and never contains spindle remnants. In many instances organelles, such as mitochondria, extend the whole length of the intercellular bridge and thus are shared by both cells (Fig. 3). Sets of double lamellae, similar to those described by Nagano (14) in bridges connecting avian meiotic spermatocytes, are occasionally present (Fig. 7). These lamellae are arranged in parallel rows which are oriented perpendicular to the long axis of the bridge.

The intercellular bridges are found between oogonia in interphase (Figs. 1 and 2), as well as between mitotic oogonia (Fig. 7). Bridges occur also between oocytes in meiosis (Fig. 8), as well as in early degeneration.

These bridges, thus, are true intercellular bridges and differ considerably from the intermediate body of Flemming (mid-body, Zwischenkörper) characteristic of mitotic telophase. However, the frequent finding of a germ cell with multiple bridges connecting it to two or more adjacent elements (Fig. 6) indicates that the bridges derive from intermediate bodies in the course of successive mitotic division in which karyokinesis is not followed by cytokinesis.

No attempts were made at this stage to correlate the incidence of the intercellular bridges with the maturation stages during the period of development studied. However, the intercellular bridges were found in significant numbers in all sections. If one considers the thinness of the sections and the fact that the bridges are obviously oriented on different planes, the actual number of these bridges must certainly be much greater than is apparent on the sections. Such a high number of cellular interconnections clearly results in the organization of the germ cells into multiple syncytial groups. The presence of organelles in the cytoplasm of the intercellular bridges indicates that material is exchanged from one cell to the other, with the result that each cell must share the developmental activity and suffer the fate of those to which it is connected. This explains not only the occurrence of mitotic, meiotic, and degenerating cells in groups but also the observation that all cells in any given group are at the same stage. The intimate cellular connections and common differentiation process also explain the close morphological similarity noted between adjacent cells and those frequently observed in the form of a mirror-image configuration (Fig. 2).

Bridges identical with those reported here have been observed by Franchi and Mandl between oogonia and oocytes of the developing rat ovary

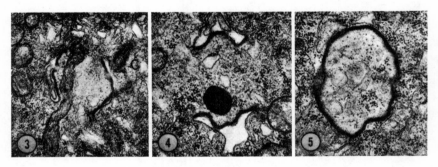

FIGURE 3 An intercellular bridge in which an elongated mitochondrion extends from one of the connected cells to the other. *C*, centriole. × 30,000.

FIGURE 4 Ribosomes, ergastoplasmic vesicles, and cytosomes in the cytoplasm of a bridge between two oogonia. × 20,500.

FIGURE 5 An intercellular bridge in cross-section showing the characteristic cylindrical configuration. Increased density of the plasma membrane permits identification of the bridge even when continuity with the connected cells is not evident. × 30,000.

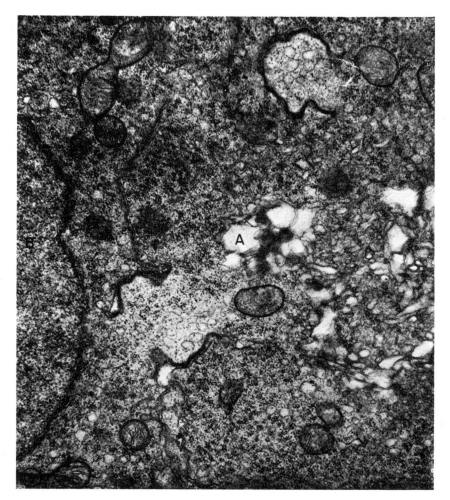

FIGURE 6 An oogonium (*A*) connected to adjacent cells by two bridges. Only one of the connected cells (*B*) is shown in the picture, since the oblique sectioning of the upper bridge prevents visualization of the other. × 25,000.

(10). Although the possible significance of this finding was not commented upon, the observation made by these authors is important in that it indicates that syncytia of germ cells may be common to all mammalian species. A syncytial organization identical with that observed by us in the ovary has been described in the seminiferous tubules of the testis by Fawcett and his collaborators (5–8). These authors have demonstrated that the spermatogenetic cells are connected by

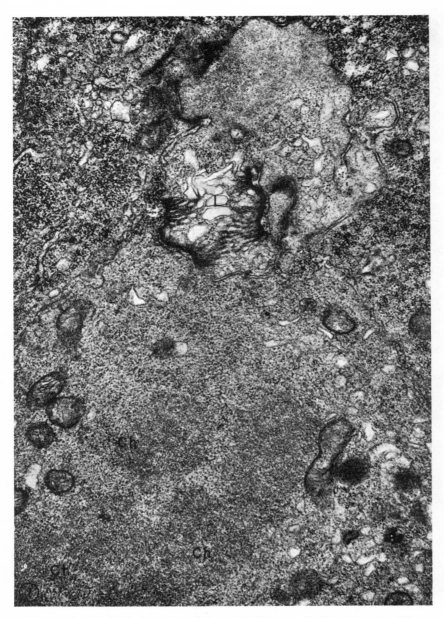

FIGURE 7 Oogonium in mitotic metaphase with intercellular bridge connecting it to another cell partially included in the section. Note the lamellar structures (*L*) which appear in the cytoplasm of the bridge and attach to the bridge plasma membrane. *Ch*, chromosomes. × 32,000.

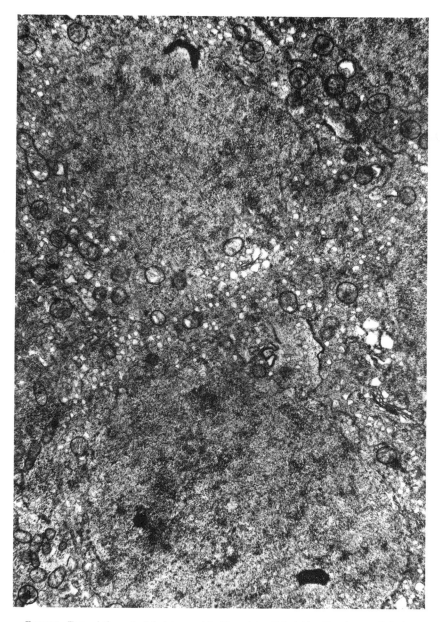

FIGURE 8 Two meiotic oocytes in leptotene conjoined by an intercellular bridge. Note the great similarity in the appearance of the two cells. × 14,000.

intercellular bridges to form cellular groups, each consisting of elements at the same stage of differentiation. These bridges, which do not include spindle remnants, were identified by Fawcett as true intercellular bridges which result in the formation of syncytia of spermatogenetic cells and represent the basis for the synchronization of cellular activity during spermatogenesis and spermiogenesis (6).

The syncytial organization of the germ cells in the developing ovary may have additional significance. The manner by which only a relatively small percentage of the total number of germ cells completes differentiation, escapes degeneration, and survives as oocytes in the follicle has long remained obscure (9). The key to survival may be complete cell division. In a relatively small number of germ cells karyokinesis may be followed by cytokinesis. These cells would then remain independent and, therefore, be able to escape the degeneration which affects the syncytial population of germ cells.

This study was supported by a Ford Foundation Grant in Reproductive Biology.

Note Added in Proof: Since the submitting of this publication there has appeared a report by Weakley on the morphology of developing germ cells in the golden hamster (1967. *J. Anat.* **101:** 435.) in which the author describes intercellular bridges between germ cells and relates them to the synchronization of germ cell differentiation.

REFERENCES

1. BAKER, T. G. 1963. *Proc. Roy. Soc. (London) Ser. B.* **158:**417.
2. BAKER, T. G. 1966. *J. Anat.* **100:**761.
3. BEAUMONT, H. M., and A. M. MANDL. 1962. *Proc. Roy. Soc. (London) Ser. B.* **155:**557.
4. BORUM, K. 1961. *Exptl. Cell Res.* **24:**495.
5. BURGOS, M. H., and D. W. FAWCETT. 1955. *J. Biophys. Biochem. Cytol.* **1:**287.
6. FAWCETT, D. W. 1961. *Exptl. Cell Res. Suppl.* **8.** 174.
7. FAWCETT, D. W., and M. H. BURGOS. 1956. Ciba Foundation Symposium Aging of Transient Tissues. Little, Brown and Company, Boston. 86.
8. FAWCETT, D. W., S. ITO, and D. SLAUTTERBACK. 1959. *J. Biophys. Biochem. Cytol.* **5:**453.
9. FRANCHI, L. L., A. M. MANDL, and S. ZUCKERMAN. 1962. Chapter 1, the development of the ovary and the process of oogenesis. *In* The Ovary. S. Zuckerman, A. M. Mandl, and P. Eckstein, editors. Academic Press Inc., London.
10. FRANCHI, L. L., and A. M. MANDL. 1962. *Proc. Roy. Soc. (London) Ser. B.* **157:**99.
11. IOANNOU, J. M. 1964. *J. Embryol. Exptl. Morphol.* **12:**673.
12. KARNOVSKY, M. J. 1961. *J. Biophys. Biochem. Cytol.* **11:**729.
13. LUFT, J. H. 1961. *J. Biophys. Biochem. Cytol.* **9:**409.
14. NAGANO, T. 1961. *Anat. Record.* **141:**73.
15. PETERS, H., E. LEVY, and M. CRONE. 1965. *J. Exptl. Zool.* **158:**169.
16. TEPLITZ, R., and S. OHNO. 1963. *Exptl. Cell Res.* **31:**183.
17. ZETTERQVIST, H. 1956. The ultrastructural organization of the columnar absorbing cells of the mouse jejunum. Ph.D. Thesis. Karolinska Institutet, Stockholm.

Morphological Changes in Mouse Eggs due to Aging in the Fallopian Tube [1]

DANIEL SZOLLOSI

Unfertilized oviductal mammalian eggs remain grossly unchanged for about 12 hours if they are not penetrated by spermatozoa. Eventually a large number of small nuclei form apparently by fragmentation of the chromosome group (Austin, '61). Details of morphological changes due to aging of eggs are of great interest because various developmental anomalies were attributed to the effects of delayed mating (Austin, '61, '67, '70; Blandau, '52; Chang, '52; German, '68; Hancock, '59; Hunter, '67; Marston and Chang, '64; Odor and Blandau, '56; Thibault, '59; Vickers, '69). A study of aging oviductal eggs was undertaken with the light and electron microscopes to serve as a baseline for further studies on eggs following delayed insemination.

MATERIALS AND METHODS

White mice of the Swiss-Webster strain were induced to ovulate by intraperitoneal injections of 5 international units (I.U.) of pregnant mare serum (PMS) (Ayerst) and 5 I.U. of human chorionic gonadotrophin (HCG) (Ayerst) given 48 hours apart. Injections were administered usually at 2 PM on the respective days. Ovulation occured 11–12 hours after the last injection in accord with the published schedule of Edwards and Gates ('59). Eggs were recovered 5, 9, 12, 14, 18 and 24 hours after estimated time of ovulation by flushing the oviducts with tissue culture medium TC 199. Oocytes were collected quickly and transferred into the fixative consisting of 1% glutaraldehyde, 0.075 M s-collidine and 0.1 M sucrose or into a 1:2 mixture of 3% glutaraldehyde solution and TC 199 tissue culture medium. The pH of the solution was re-adjusted to neutrality with a few drops of dilute NaOH solution. Subsequently the eggs were postosmicated in an 1% osmium tetroxide solution in s-collidine buffer. The eggs were dehydrated in alcohol, passed through mixtures of absolute alcohol and Epon 812 embedding medium with increasing concentrations and finally embedded in Epon 812 (Luft, '61). Sections were doubly stained with saturated uranyl acetate and lead citrate (Venable and Coggeshall, '65) and examined in a Philips 200 electron microscope. For light microscopy 0.5 μ sections were stained with Richardson's ('60) basic methyl blue-azure II stain and photographed with phase optics.

OBSERVATIONS

Eggs 5–12 hours in the Fallopian tube. No microscopically detectable changes are observed in oviductal oocytes up to 12 hours after ovulation. The eggs are arrested in metaphase of meiosis two (fig. 1).

---
[1] Supported by the National Institutes of Health research grant HD 03752 and United States Public Health Service research grant GM 16598.

The spindle is located peripherally and its axis is oriented tangentially to the egg surface as observed in most freshly ovulated mammalian eggs (for a review, see Austin, '61). Microvilli are distributed over the entire egg surface with the exception of the region overlying the spindle (figs. 2, 9). A filament bundle usually forms a core in the microvilli, while over the spindle a layer of filaments runs parallel to the egg cell membrane (fig. 6) (Szollosi, '70a). Cortical granules are distributed in irregular clusters throughout the peripheral zone of the oocyte and are usually a few hundred Ångstroms from the inner leaflet of the egg cell membrane (fig. 2) (Szollosi, '67) with the exception of the region overlying the spindle.

The spindle or the achromatic figure is constituted almost exclusively of microtubules and chromosomes. The microtubules are attached to the kinetochores in bundles (Szollosi, '70b). Most other cell organelles are excluded from this region; occasionally clusters of polysomes and small vesicles are seen (fig. 1). In no instance was a centriole located at the spindle poles even though several spindles were sectioned serially. Bundles of microtubules focus on a cluster of filamentous aggregations, structurally similar to diffuse centriolar satellites or pericentriolar bodies (fig. 3) (Bernhard and de Harven, '60; Szollosi, '64; de The, '64). The filamentous material of the pericentriolar bodies loosely interlinks the foci. A group of granules is located consistently near the spindle poles. The granules range between 500–1200 Å in diameter and they vary in electron density (fig. 10). Similar granules are present in small clusters in other regions of unfertilized eggs.

The spindle is surrounded by cytoplasmic organelles. Filamentous material (Enders and Schlafke, '65; Weakley, '66) is the most frequent component. It has a specific periodicity and may represent yolk deposition (Szollosi, '65b). This structure is found particularly in oocytes and during cleavage stages. Mitochondria, individual vesicles, smooth tubules and vesicular complexes of varying dimensions are often found proximal to the spindle. Occasionally a ribosome-filament complex is located near the chromosomes or in the periphery of the spindle (Szollosi, '70b).

Eggs 14–18 hours in the Fallopian tube. The position of the spindle is the most conspicuous morphological change in these eggs. Light micrographs disclose that the spindle rotates 90° and the spindle axis becomes disposed radially. One spindle pole remains in the proximity of the egg surface (fig. 4). In some eggs the spindle separates from the surface completely and appears to migrate part-way into the ooplasm (fig. 5). The relative position of the spindle and its orientation can be determined only with difficulty. In most cases serial thick and thin sections were examined to overcome this disadvantage. On the fine structural level the spindle changes little. During rotation small cytoplasmic organelles such as elements of the filamentous material and small vesicles intermingle with the spindle fibers. Some chromosomes may be separated from the metaphase plate in a few eggs 14–18 hours after reaching the oviduct, apparently due to partial destruction of the spindle (fig. 6). Only those chromosomes which remain at the metaphase plate are distinctly associated with spindle microtubular elements. In certain stretches of the egg surface the cortical granules move to greater depths in the egg cytoplasm (fig. 9). No changes are detected in the egg surface itself.

Between 12 and 18 hours most ovulated eggs are devoid of cumulus cells. Cytoplasmic remnants representing the terminal enlargements of the processes of the corona radiata are seen frequently in the perivitelline space. The remaining follicle cells though morphologically intact may be in varying stages of degeneration.

Eggs 18–24 hours in the Fallopian tube. The meiotic spindle now lies near the geometric center of the egg, the usual position of the mitotic spindle (fig. 7). The centrally located spindle is penetrated regularly by a variety of cytoplasmic organelles, e.g., filamentous material, vesicles of various dimensions, mitochondria and vesicular complexes (fig. 10). A special group of granules with variable electron densities still cluster near the spindle pole. They apparently migrate along with the spindle as it rotates and moves more centrally. In some cases the chromosomes dissociate from the spindle; they unwind to varying degrees and a clear layer surrounds the chromosome set indicating an attempt at reconstitution of the nucleus (fig. 8). In others a large nucleus is formed containing several nucleoli (fig. 11). The nucleoli are formed exclusively of filamentous components (fig. 13) similar to those described in pronuclei of fertilized eggs (Szollosi, '70b). Only a single nucleus is formed in eggs in which the nuclei are reconstituted. The nuclear envelope limiting the chromatin possesses a large number of nuclear pores. Occasional "blebs" occur along the nuclear envelope (fig. 13, inset) (Szollosi, '65, '70b). The second polar body is not formed in these eggs. Some aged eggs fragment into several pieces. The largest cytoplasmic fragment contains one large nucleus, identical to that described above. No nuclear material is found in the smaller cytoplasmic fragments (fig. 12). Fragmentation appears to involve cytoplasmic constrictions similar to cell cleavage. A filamentous system can be observed at the edge of the constrictions. The filaments are positioned parallel to the plane of the partial furrows (Szollosi, '70a).

In some eggs large lysosome-like granules develop in the egg cytoplasm (fig. 14). Several granules form large clusters near the vesicular complexes. The individual granules range between 0.3 and 2.0 μ in diameter. The smaller granules are located in the vesciular complex and seem to develop within the vesicles while the larger ones are more peripherally located. As observed with the light microscope the granules stain deeply with the alkaline methyl blue-azure II stain (Richardson et al., '60).

Morphological changes are also found in the structure of the cortical granules after 18–24 hours of aging. As the granules move more centrally several of them swell. The contents of the swollen granules do not disperse uniformly within their limiting membranes. Local condensations of the granule content form an intricate pattern. In swollen granules the filamentous nature of the cortical granule content becomes clearly recognizable (fig. 15).

DISCUSSION

Ovulated mouse eggs aged in the oviducts apparently do not undergo any morphological changes during the first 12 hours. The eggs are not activated spontaneously. Morphological changes are observed in unfertilized eggs residing in the Fallopian tubes for 14 hours.

Rotation and migration of the spindle to the central portion of the ovum are the first and most obvious changes. The central portion of the spindle would potentially prepare the egg for equal cleavage rather than the usual unequal division which forms the second polar body and the ovum. Only one polar body was found in every egg examined. Recently two reports were published on experimental parthenogenesis in the mouse. Electric shock applied to the exposed oviduct (Tarkowski et al., '70) and hyaluronidase treatment in a hypotonic medium (Graham, '70) were used respectively to activate the eggs. Following both procedures, development proceeded to the blastocyst stage. In the former case, eggs were activated two to four hours after ovulation while in the latter case the eggs were 12 to 17 hours old. In my study, eggs aged 12–17 hours in the oviduct regularly revealed that the spindle had rotated and, in a few cases, begun to migrate to a more central position at the time when in Graham's study ('70) hyaluronidase was employed as a parthenogenetic agent. It could be argued that separation of the spindle from the cell membrane and its central migration were accelerated by the electric shock (Tarkowski et al., '70); in this case cleavage would give rise to two haploid cells instead of an ootid and a polar body. Such a division produced with physical and/or chemical means is similar to "immediate cleavage" after delayed fertilization (Braden and Austin, '54b).

In the present experiments the eggs might develop along two independent pathways. If they divide equally without DNA synthesis, haploid embryos may be formed, a situation identical to that discussed above. Alternatively, eggs could form a single nucleus similar to functional pronuclei but encompassing the paired homologous chromosomes. The homologous chromosomes would separate and an interphase-like nucleus be formed. At the same time the diploid number would be reconstituted. Such a nucleus would undergo DNA synthesis in a manner similar to pronuclei of fertilized mammalian eggs (Oprescu and Thibault, '65; Szollosi, '66). The reconstituted nuclei cannot be distinguished morphologically from pronuclei or nuclei of blastomeres of a two-cell stage. The appearance of several nucleoli and formation of nuclear blebs suggests that nuclei are functional (Szollosi, '65a).

Cellular organelles usually do not penetrate either meiotic or mitotic spindles. However, in aging eggs various small cytoplasmic organelles penetrate the second meiotic spindle as a result of its rotation and migration. This observation is in agreement with the view of Hiramoto ('69) that the visco-elastic properties of fertilized eggs correlate closely with the structural elements of the spindle. The viscosity of the hyaloplasm apparently remains low and the small cell structures, e.g., filamentous material, vesicles, polysomes, can penetrate freely between the microtubule bundles. Occasionally even large organelles such as mitochondria invade the spindle area. In most eggs, however, the structural integrity of the spindle was retained in contrast to the findings of others (Marston and Chang, '64).

The absence of centrioles at the spindle poles similar to most cells of higher plants demonstrates that these are not essential cell organelles for the formation of the spindle apparatus in animal cells nor are they necessary for cell division. However, the structural elements, i.e., the continuous and kinetochore fibers, are present in the meiotic spindle of the mouse eggs studied. Microtubules project toward nucleation centers; these are morphologically similar to diffuse centriolar satellites (Bernhard and de Harven, '60; Szollosi, '64). The nucleation center represents a large area of thin filamentous components within which smaller condensations (satellites) represent the origin or termination of microtubules. The individual satellites apparently can be displaced from the nucleation center causing separation of one or several of the chromosomes from the metaphase plate. The same result would be obtained if the microtubules of certain chromosomes were disintegrated. The formation of micronuclei or nuclear fragments may depend on the capability of such separated chromosomes to reconstitute a nuclear envelope. Nucleoli form in most micronuclei. Micronuclei containing nucleoli may be induced by colcemid treatment (Edwards, '61). The presence of the cluster of granules with electron-dense contents at the spindle poles is a new observation. The granules remain at the spindle poles even though the spindle rotates and migrates. The role of the vesicles is not known. They may be related to the formation of spindle microtubules. Similar vesicles have been observed at the spindle regions during formation of the first meiotic spindle in mouse eggs (Calarco, Szollosi and Donahue, unpublished).

We have clearly established in this study that in the aging egg the cortical granules migrate away from the cell membrane and into the central ooplasm. The number of granules remaining at the egg cortex is inversely proportional to the age of the eggs. The reports of increase in polyspermy as eggs are aged (Braden and Austin, '54a; Odor and Blandau, '56; Marston and Chang, '64; Hunter, '67), further implicate the cortical granules in the defense mechanism against multiple fecundation (Szollosi, '67). Cortical granules are apparently unstable organelles. With aging several swell and an intricate pattern becomes evident within their limiting membrane. The pattern observed may be caused by failure of the granule content to disperse uniformly after swelling. It is possible, however, that in the mouse, compaction of some of the cortical granules is not completed at the time of ovulation. In hamster eggs the cortical granules may be extruded into the perivitelline space (Yanagimachi and Chang, '61). As the periodic acid-Schiff-positive material, presumably cortical granule content, increases in the perivitelline space with aging, fewer spermatozoa penetrate the zona pellucida. In aging mouse eggs no evidence was found for the extrusion of cortical granules as would occur at the time of normal sperm penetration (Szollosi, '67 and unpublished).

Degenerative changes have been noted in the cytoplasm of many eggs aged for

18 to 24 hours after ovulation. We interpret the large, electron-dense granules to be autophagic vacuoles. Their proximity to the large vesicular complexes may further indicate that they belong to the lysosome family. No cytochemical tests were performed, however, to confirm these assumptions.

The cell surface does not appear to change during the aging process. Irregularly arranged microvilli cover the entire egg surface. The egg membrane overlying the second maturation spindle lacks microvilli and cortical granules. The arrangement of the cortical filaments suggests that this site represents the cell surface where the second polar body was recently extruded. Cortical granules from that cytoplasmic region become incorporated into the first polar body (Szollosi, unpublished).

Fragmentation of eggs at the time of the first cleavage (20–24 hours after estimated time of sperm entry) is interesting since the fragmentation is similar to cytokinesis in detail. At the point of abortive "furrowing," thin filaments are oriented in the plane of the furrow (Szollosi, '70b). In every case observed the large nucleus remained intact and no chromosomal material nor nuclear fragments were found in the smaller cytoplasmic blebs.

ACKNOWLEDGMENTS

Many thanks are due to Dr. Patricia Calarco for stimulating discussions on mouse egg development and for her critical comments regarding the manuscript. The author also thanks Dr. Richard Blandau for reviewing the manuscript.

LITERATURE CITED

Austin, C. R. 1961 The Mammalian Egg. Blackwell Sci. Pub., Oxford.

——— 1967 Chromosome deterioration in aging eggs of the rabbit. Nature, 213: 1018–1019.

——— 1970 Ageing and reproduction: postovulatory deterioration of the egg. J. Reprod. Fertil., 12 (Suppl.): 39–53.

Austin, C. R., and A. W. H. Braden 1953 An investigation of polyspermy in the rat and rabbit. Aust. J. Biol. Sci., 6: 674–692.

Bernhard, W., and E. de Harven 1960 L'ultrastructure de centriole et d'austres éléments de l'appareil achromatique. 4th Internat. Conf. Electron Microscopy, Berlin (1958), 218.

Blandau, R. J. 1952 The female factor in fertility and infertility. I. Effects of delayed fertility on the development of the pronuclei in rat ova. Fertil. Steril., 3: 349–365.

Braden, A. W. H., and C. R. Austin 1954a Fertilization of the mouse egg and the effect of delayed coitus and of hot shock treatment. Aust. J. Biol. Sci., 7: 552–565.

——— 1954b Reaction of unfertilized mouse eggs to some experimental stimuli. Exp. Cell Res., 7: 227–280.

Chang, M. C. 1952 Effects of delayed fertilization on segmenting ova, blastocysts and fetuses in rabbit. Fed. Proc., 11: 24.

de The, G. 1964 Cytoplasmic microtubules in different animal cells. J. Cell Biol., 23: 265–275.

Edwards, R. G. 1961 Induced heteroploidy in mice: effect of deacetylmethylcolchicine on eggs at fertilization. Exp. Cell Res., 24: 615–617.

Edwards, R. G., and A. H. Gates 1959 Timing of the stages of the maturation division, ovulation, fertilization and the first cleavage of eggs of adult mice treated with gonadotrophins. J. Endocr., 18: 292–304.

Enders, A. C., and S. J. Schlafke 1965 The fine structure of the blastocyst: some comparative studies. In: Preimplantation Stages of Pregnancy. Ciba Found. Symp. G. E. W. Wolstenholm, ed. Little, Brown and Company, Boston, 29–54.

German, J. 1968 Mongolism, delayed fertilization and human sexual behavior. Nature, 217: 516–518.

Graham, C. F. 1970 Parthenogenetic mouse blastocysts. Nature, 226: 165–167.

Hancock, J. L. 1959 Polyspermy of pig ova. Anim. Prod., 1: 103–106.

Hiramoto, Y. 1969 Mechanical properties of the protoplasm of the sea urchin egg. III. Fertilized egg. Exp. Cell Res., 56: 209–218.

Hunter, R. H. F. 1967 The effects of delayed insemination on fertilization and early cleavage in the pig. J. Reprod. Fertil., 13: 133–147.

Luft, J. H. 1961 Improvements in epoxy embedding. J. Biophys. Biochem. Cytol., 9: 409–414.

Marston, J. H., and M. C. Chang 1964 The fertilizable life of ova and their morphology following delayed insemination in mature and immature mice. J. Exp. Zool., 155: 237-252.

Odor, D. L., and R. J. Blandau 1956 Incidence of polyspermy in normal and delayed mating in rats of the Wistar strain. Fertil. Steril., 7: 456–457.

Oprescu, St., and C. Thibault 1965 Duplication of DNA in the egg of the rabbit after fertilization. Ann. Biol. Anim. Biochem. Biophys., 5: 151–156.

Richardson, K. C., L. Jarett and E. H. Finke 1960 Embedding in epoxy resin for ultrathin sectioning in electron microscopy. Stain Tech., 35: 313–323.

Szollosi, D. 1964 Structure and function of centrioles and their satellites from the jellyfish, *Phialidium gregarium*. J. Cell Biol., 21: 465–479.

———— 1965a Extrusion of nucleoli from pronuclei of the rat. J. Cell Biol., 25: 545–562.

———— 1965b Development of "yolky substance" in some rodent eggs. Anat. Rec., 151: 424.

———— 1966 Time and duration of DNA synthesis in rabbit eggs after sperm penetration. Anat. Rec., 154: 209–212.

———— 1967 Development of cortical granules and the cortical reaction in rat and hamster eggs. Anat. Rec., 159: 431–446.

———— 1970a Cortical cytoplasmic filaments of cleaving eggs: a structural element corresponding to the contractile ring. J. Cell Biol., 44: 192–210.

———— 1970b Nucleoli and ribonucleoprotein particles in the preimplantation conceptus of the rat and mouse. In: The Biology of the Blastocyst. R. J. Blandau, ed. Univ. of Chicago Press, Chicago.

Tarkowski, A. K., A. Witkowska and J. Nowicka 1970 Experimental parthenogenesis in the mouse. Nature, 226: 162–165.

Thibault, C. 1959 Analyse de la fecondation de l'oeuf de la truie après accouplement ou insémination artificielle. Ann. Inst. Nat. Rech. Agron., Paris, Ser. D., Suppl., pp. 165–177.

Venable, J. H., and R. Coggeshall 1965 A simplified lead citrate stain for use in electron microscopy. J. Cell Biol., 25: 407–408.

Vickers, A. D. 1969 Delayed fertilization and chromosomal anomalies in mouse embryos. J. Reprod. Fertil., 20: 69–76.

Weakley, B. 1966 Electron microscopy of the oocyte and granulosa cells in the developing ovarian follicles of the golden hamster (*Mesocricotus auratus*). J. Anat., 100: 503–534.

Yanagimachi, R., and M. C. Chang 1961 Fertilizable life of golden hamster ova and their morphological changes at the time of losing fertilizability. J. Exp. Zool., 148: 185–204.

PLATE 1

EXPLANATION OF FIGURES

1   The second maturation spindle of a mouse egg recovered approximately five hours after ovulation lies near the cell surface. The spindle area is relatively free of cytoplasmic organelles except for occasional small vesicles and polysomes. Microvilli are absent from the cell surface overlying the spindle. A thin layer of filaments (f) lies parallel to the cell membrane. Cortical granules are absent from this region of the cell surface. Near the chromosomes a vesicular complex (vc) can be recognized. × 9,400.

2   The egg surface of a recently ovulated egg demonstrates microvilli and numerous cortical granules. (CG). × 27,000.

PLATE 1

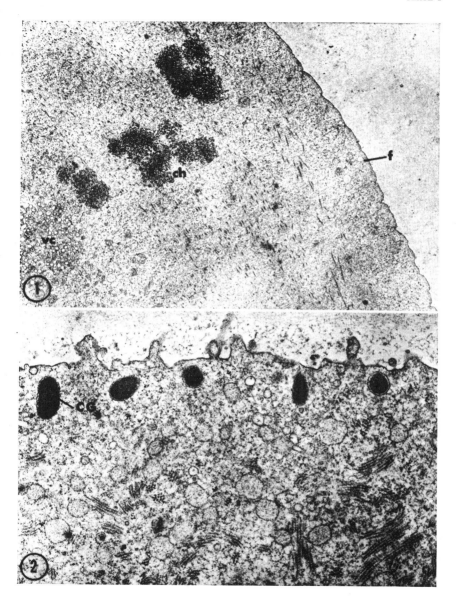

PLATE 2

EXPLANATION OF FIGURES

3 At the spindle pole microtubules associated with aggregations of thin filamentous material similar to satellites (S) or pericentriolar bodies. ×45,000.

4 The second maturation spindle rotates and takes up a radial position in eggs which spent 14–18 hours in the Fallopian tube. Most cells of the corona radiata have fallen off the zona pellucida (zp). ×600.

5 The spindle moves more centrally in other eggs at 14–18 hours after ovulation. ×600.

PLATE 2

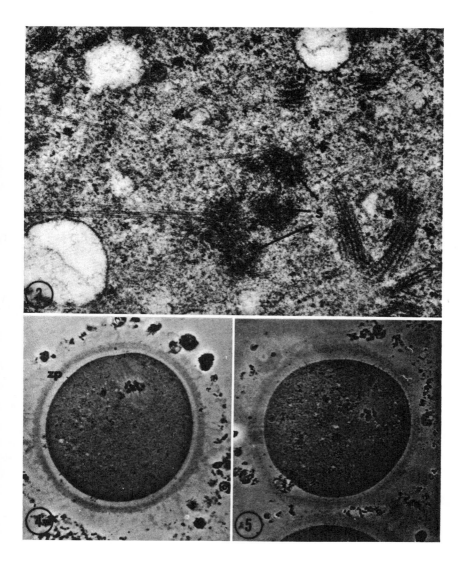

PLATE 3

EXPLANATION OF FIGURES

6 Some chromosomes (ch*) apparently separate from the metaphase plate at 14–18 hours after ovulation. Microtubules are not recognized near the separated chromosome. Other chromosomes (ch) are among spindle microtubules (m). Parallel to the smooth portion of the cell membrane a layer of filaments (f) is observed. ×7,300.

7 The spindle moves to the approximate geometric center of unfertilized eggs 18–24 hours after ovulation. ×600.

8 The spindle breaks down and the chromatin partially unwinds in an attempt to reconstitute a nucleus in some eggs recovered 18–24 hours after ovulation. × 600.

PLATE 3

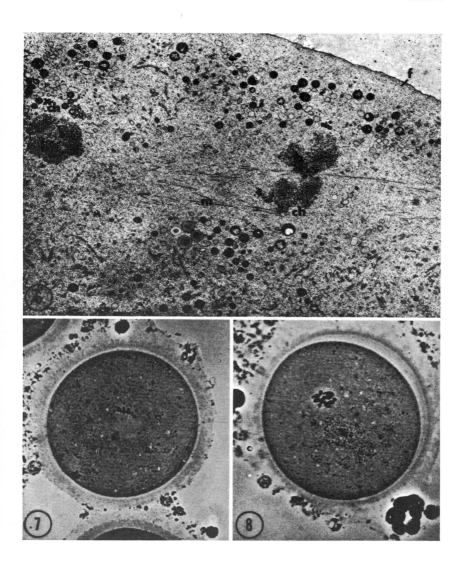

PLATE 4

EXPLANATION OF FIGURES

9 The cortical granules (CG) migrate centrally from the egg surface in oviductal eggs aged 18–24 hours. Microvilli with a filamentous core cover the largest portion of the egg surface. ×15,500.

10 A portion of a spindle which migrated to the cell center in an egg recovered 18–24 hours after ovulation. Several clusters of filamentous material (fm) and vesicles of various diameters are seen within the spindle. Near the spindle pole a group of vesicles is found with a mottled electron-dense content (g). Ch, chromosomes; m, microtubules. ×15,500.

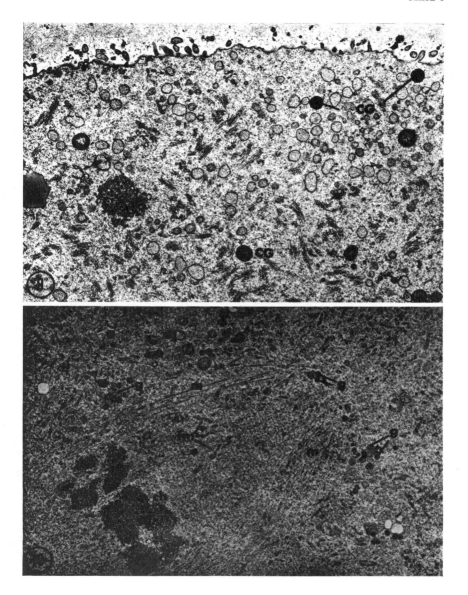

PLATE 5

EXPLANATION OF FIGURES

11 The nucleus is reconstituted in several eggs. Four nucleoli are seen in this nucleus from an egg recovered 18–24 hours after ovulation. ×600.

12 Few fragmenting eggs are recovered 18 to 24 hours after ovulation. The largest cytoplasmic fragment contains a large nucleus with several nucleoli. Micronuclei are lacking from the smaller cytoplasmic fragments. ×600.

13 An electron micrograph of a reconstituted nucleus shows that the nuclear envelope has reformed. A few nuclear blebs can be found occasionally along the envelope (b). The nucleolus (nu) is constituted only of thin filamentous material. ×10,500. Inset, ×28,000.

PLATE 5

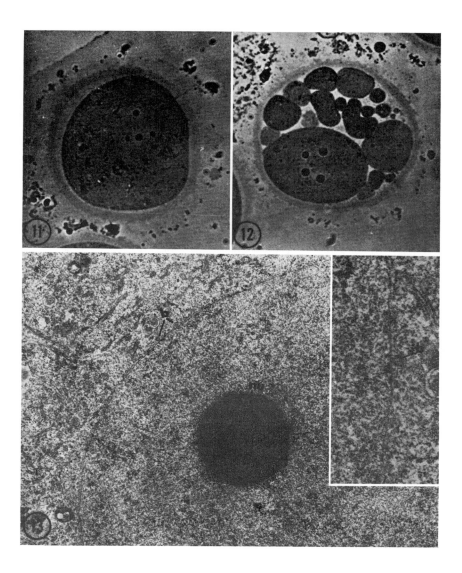

PLATE 6

EXPLANATION OF FIGURES

14 Near the vesicular complexes lysosome-like granules (ly) develop occasionally with different electron densities in eggs aged 18–24 hours. ×11,000.

15 The cortical granules (CG) swell in eggs aged 18–24 hours as they migrate centrally. Their contents swell unevenly and form an intricate pattern. ×23,000.

PLATE 6

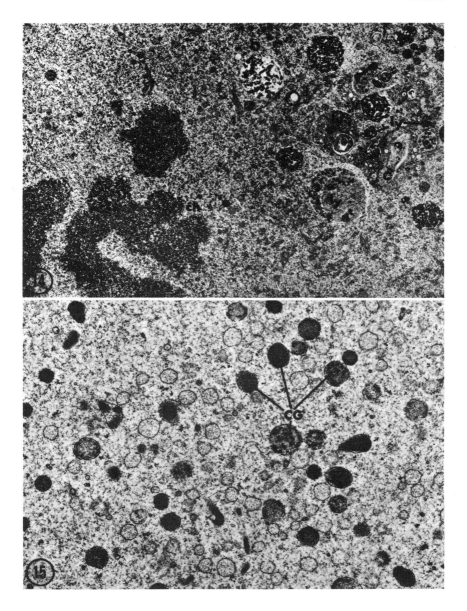

Genetic Effects on Meiosis

GENETIC ANALYSIS OF EIGHT-SPORED ASCI PRODUCED BY GENE E IN NEUROSPORA TETRASPERMA[1]

FORD CALHOUN[2] AND H. BRANCH HOWE, JR.

UNLIKE the better-known eight-spored species of Neurospora, *N. tetrasperma* usually produces asci with only four ascospores; each ascospore contains nuclei derived from two of the four meiotic products and is usually heterokaryotic for the mating type alleles. DODGE (1939) showed that eight-spored asci also occurred in *N. tetrasperma*, in crosses heterozygous for the pleiotropic gene *E*. Such crosses often showed a high frequency of aborted asci but when fertile yielded varying proportions of both eight-spored and four-spored asci in the same perithecia. Crosses homozygous for *E* were never fertile. The vegetative cycle was also affected, for ascospores containing the *E* allele usually died as germlings, although their survival frequency could be somewhat increased by enriched media.

The cytology of asci heterozygous for *E* has been previously investigated (DODGE, SINGLETON and ROLNICK 1950; SINGLETON, cited by BRAVER 1952), but these authors, as well as HOWE and HAYSMAN (1966), made only limited genetic analyses. Both the interest and the potential usefulness of *E* made it desirable to gain a further understanding of the nature of this allele. The present analysis of asci heterozygous for *E* aims to interpret the various segregation patterns and to correlate these patterns with earlier cytological observations. Our efforts were greatly facilitated by using one of DODGE's *E* strains which was found to have lost much of the lethality described above, such that ascus abortion was markedly reduced, and all eight ascospores from the asci usually survived the germling stage.

MATERIALS AND METHODS

The eight mutants of *N. tetrasperma* used were $act(113)$, $al(102)$, $col(105)$, $col(118)$, $col(120)$, $E(121)$, $me(123)$, and $pan(124)$. All of these mutants, as well as the culture media used, have been described previously (HOWE and HAYSMAN 1966).

Because the asci remained optimal for dissection for only very short periods, dissections were usually done about 10 days after conidial subculture of a previously established cross. Crosses were maintained at room temperature (23–28°C). All asci were dissected and isolated serially, but without distinction between the apex and base, except where specified. To improve germination the isolated ascospores were aged under conditions of high humidity, so as to avoid drying of the slants, prior to heat activation. Ascospore viability of about 68% was obtained after aging for at least three weeks at room temperature.

[1] This work was supported by Public Health Service Grant GM 10672.
[2] Supported in part by Public Health Service Training Grant 5-T1-GM-767-05.

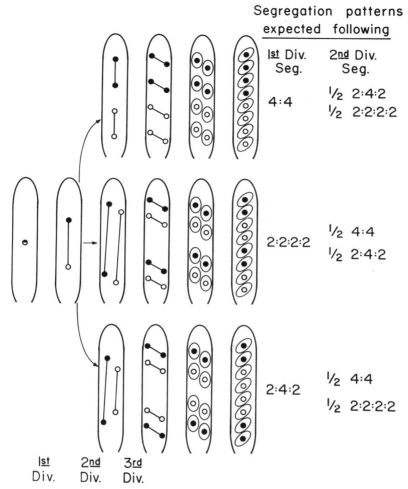

FIGURE 1.—Diagram of the development of eight-spored asci heterozygous for E in *N. tetrasperma* (adapted from other authors; see text) to account for the three segregation patterns observed genetically in the present study. To the right are the patterns expected following first division segregation (shown in diagram) and second division segregation of an allelic-pair. *Upper Row, Second Division:* Tandem spindles and no nuclear passing. *Middle Row, Second Division:* Overlapping spindles and passing of a nucleus beyond a nonsister nucleus. *Lower Row, Second Division:* Overlapping spindles and passing of a nucleus beyond two nonsister nuclei.

The E allele was scored by crossing each isolate from every ascus to both EA and Ea tester stocks. All isolates containing the E allele were identified by failure to show fertility with either tester, since $E \times E$ crosses are infertile.

Corrected chi-square values were obtained by YATES' correction.

Terminology: The terms concordance and discordance refer to the two members of an ascospore-pair (sister ascospores) being adjacent or nonadjacent, respectively. Asci having the two members of all four ascospore-pairs adjacent are defined as concordant; asci with the two members of at least one ascospore-pair nonadjacent are defined as discordant. Asci are defined as tetratype if at least two segregating allelic-pairs show the tetratype relationship. Ditype asci are defined as those having all segregating allelic-pairs in the ditype relationship. This terminology is used only with eight-spored asci.

RESULTS AND DISCUSSION

Comparison of cytological and genetic studies: The cytology of asci heterozygous for E revealed two types of spindle orientations at the second meiotic division (DODGE, SINGLETON and ROLNICK 1950; SINGLETON, cited by BRAVER 1952). In the more frequent type, the two spindles overlap leading to nuclear passing and the consequent location of two nonsister nuclei at each end. In the less frequent type, the two spindles are in tandem resulting in the even spacing of the four nuclei along the longitudinal axis of the ascus and the consequent location of two sister nuclei at each end.

The present genetic study reveals that the mating type alleles, although known to segregate consistently with the centromere at the first meiotic division (HOWE and HAYSMAN 1966), nevertheless show three different segregation patterns, 4:4, 2:2:2:2, and 2:4:2, in serially dissected asci. We propose that these three different segregation patterns of the centromere marker may be reconciled in the following manner with the two different spindle orientations just described. The 4:4 pattern probably arises from the infrequent tandemly oriented second division spindles (Figure 1, upper row). The 2:2:2:2 and 2:4:2 patterns probably arise from the more frequent overlapping second division spindles; each of these latter two patterns, although genetically distinguishable, results from different degrees of nuclear passing (Figure 1, middle and lower rows), which probably would not be cytologically distinguishable, and which were, therefore, reported as only one type of event by SINGLETON. It will be shown that our observed frequencies of segregation patterns are reasonably consistent with the frequencies SINGLETON reported for the two different types of spindle orientations. The segregation patterns expected for markers segregating at the second division, assuming the same cytogenetic mechanisms just proposed, are also presented in Figure 1.

Summation of asci analyzed: A total of 515 asci was dissected and isolated serially from five different crosses heterozygous for E. Ten of the 515 asci had allelic ratios other than 1:1 for one or more allelic-pairs and are not considered further; complete genotypes of all 515 asci are given elsewhere (CALHOUN 1968). Ascospore viability in the remaining 505 asci was as follows:

\	\	\	Number of viable ascospores per ascus						Total
0	1	2	3	4	5	6	7	8	asci
77	17	11	17	50	18	42	82	191	505

Our analysis is based upon the 273 asci that contained at least seven viable ascospores, which is the minimum viability needed to attempt an unbiased interpretation of ascospore arrangements. A total of 125 of the 273 asci (45.8%) was concordant; 148 (54.2%), discordant. Results and interpretations for these two groups will be presented separately.

Concordant asci: There were 79 ditypes and 46 tetratypes among the 125 concordant asci. Each of these two ascus types is further grouped according to the segregation patterns shown by the mating type alleles (Table 1). Since mating type segregates consistently at the first division, a second marker showing the same segregation pattern may be assumed to have segregated also at the first

TABLE 1

Genotypes of 125 concordant eight-spored asci serially dissected from five different crosses

Cross	Segregation pattern of mating type alleles	Ascus type*	Number of asci	Genotype of ascospore-pairs			
1. +Ea × al+A	4:4	D	1	+Ea	+Ea	al+A	al+A
	2:2:2:2	D	4	alEA	++a	alEA	++a
		D	1	al+A	+Ea	al+A	+Ea
		T	1	alEA	++a	al+A	+Ea
	2:4:2	D	1	al+A	+Ea	+Ea	al+A
2. ++Ea × act(113)col(118)+A	4:4	D	2	++Ea	++Ea	mc+A	mc+A
(act=m in these 23 asci to		D	1	mc+a	mc+a	++EA	++EA
avoid confusion with a		D	2	mcEA	mcEA	+++a	+++a
mating type allele)		T	1	mcEA	mc+A	++Ea	+++a
	2:2:2:2	D	3	mc+A	++Ea	mc+A	++Ea
		D	6	mc+a	++EA	mc+a	++EA
		D	1	+++A	mcEa	+++A	mcEa
		T	1	+++a	mcEA	++Ea	mc+A
		T	1	mc+a	++EA	mcEa	+++A
		T	1	mc+a	+cEA	+++a	m+EA
		T	1	+++A	++Ea	mc+A	mcEa
		T	1	mc+A	mcEa	+++A	++Ea
	2:4:2	D	1	++Ea	mc+A	mc+A	++Ea
		D	1	+++a	mcEA	mcEA	+++a
3. +pan EA × col(120)++a	4:4	D	2	+pEA	+pEA	c++a	c++a
		D	1	c+EA	c+EA	+p+a	+p+a
		D	3	c++A	c++A	+pEa	+pEa
		D	1	c+Ea	c+Ea	+p+A	+p+A
		T	1	cp+a	+p+a	++EA	c+EA
		T	1	+p+a	cp+a	c+EA	++EA
		T	1	+pEA	c+EA	+++a	cp+a
	2:2:2:2	D	7	c++A	+pEa	c++A	+pEa
		D	4	+pEA	c++a	+pEA	c++a
		D	4	+p+A	c+Ea	+p+A	c+Ea
		D	4	+p+a	c+EA	+p+a	c+EA
		D	1	cpEA	+++a	cpEA	+++a
		T	1	++EA	cp+a	c+EA	+p+a

133

TABLE 1—Continued

Genotypes of 125 concordant eight-spored asci serially dissected from five different crosses

Cross	Segregation pattern of mating type alleles	Ascus type*	Number of asci	Genotype of ascospore-pairs			
4. +pan EA × col(105)++a	4:4	T	1	+pEA	++·+a	cpEA	c++a
		T	1	+++A	cpEa	c++A	+pEa
		T	1	+p+A	c+Ea	cp+A	++Ea
		T	1	+pEA	c++a	cpEA	+++a
		T	1	+p+a	c+EA	c++a	+pEA
		D	1	c++a	c++a	+pEA	+pEA
		D	2	+pEa	+pEa	c++A	c++A
		D	1	c+Ea	c+Ea	+p+A	+p+A
		T	1	+++A	c++A	cpEa	+pEa
		T	1	c++a	+++a	+pEA	cpEA
		T	1	+++a	c++a	+pEA	cpEA
		T	1	+p+a	cp+a	c+EA	++EA
		T	1	cpEa	+pEa	+++A	c++A
		T	2	++Ea	c+Ea	cp+A	+p+A
	2:2:2:2	D	3	c++A	+pEa	c++A	+pEa
		D	5	+p+a	c+EA	+p+a	c+EA
		D	2	c++a	+pEA	c++a	+pEA
		D	5	+p+A	c+Ea	+p+A	c+Ea
		D	1	cpEa	+++A	cpEa	+++A
		T	2	c+EA	cp+a	++EA	+p+a
		T	2	c++a	cpEA	+++a	+pEA
		T	1	cpEa	c++A	+pEa	+++A
		T	1	+p+A	cpEa	c++A	++Ea
		T	1	+p+A	c+Ea	cp+A	++Ea
		T	2	+pEA	c++a	cpEA	+++a
		T	1	cpEa	+++A	+pEa	c++A
		T	1	c++A	cpEa	+++A	+pEa
		T	1	cp+A	c+Ea	+p+A	++Ea
	2:4:2	D	1	c++A	+pEa	+pEa	c++A
		T	1	+p+A	c+Ea	++Ea	cp+A
		T	1	c++A	+p+a	+pEa	c+EA
		T	1	cp+a	c+EA	++EA	+p+a
5. +Ea × me+A	4:4	D	2	mEA	mEA	++a	++a
		D	1	++A	++A	mEa	mEa
	2:2:2:2	D	2	m+A	+Ea	m+A	+Ea
		D	1	m+a	+EA	m+a	+EA
		D	1	++A	mEa	++A	mEa
		T	1	++a	mEA	m+a	+EA
		T	3	m+a	mEA	++a	+EA
		T	1	mEa	++A	+Ea	m+A
		T	1	m+a	+EA	++a	mEA
		T	1	mEa	m+A	+Ea	++A
	2:4:2	T	1	mEa	++A	m+A	+Ea

* D signifies ditype with respect to all segregating pairs of alleles; T signifies tetratype with respect to at least two segregating pairs of alleles.

TABLE 2

Frequencies of the three segregation patterns for the mating type alleles in 125 concordant ditype (D) and tetratype (T) asci

Segregation pattern	Cross and ascus type										Total asci		
	1		2		3		4		5				
	D	T	D	T	D	T	D	T	D	T	D	T	Percent
4:4	1	0	5	1	7	3	4	7	3	0	20	11	24.8
2:2:2:2	5	1	10	5	20	6	16	12	4	7	55	31	68.8
2:4:2	1	0	2	0	0	0	1	3	0	1	4	4	6.4
χ^2_1*	0.38		1.17		0.18		2.22		1.76		0.07		

* Chi-square tests show that segregation patterns are independent of ascus types.

division; a second marker showing a different segregation pattern than mating type may be assumed to have segregated at the second division. It may be seen from the groupings in Table 1 that three different segregation patterns for mating type occurred and that the other markers in any given ascus sometimes showed the same pattern as mating type and sometimes did not. Consequently, all three patterns occurred following first division segregation, and all three occurred following second division segregation. Centromere markers have been used similarly in other studies to classify segregations of markers in asci having uncertain ascospore arrangements (HAWTHORNE and MORTIMER 1960; BERG 1966).

In Table 2, chi-square values from contingency tables are shown which indicate independence of segregation patterns and ascus types (ditypes and tetratypes). That is, none of the three segregation patterns was any more likely to occur in one ascus type than in the other. Consequently, discordance did not occur in significant frequencies at the ends of the 4:4 ditype asci nor at the center of the 2:4:2 ditype asci; thus discordant ditype asci were not misclassified as concordant asci with significant frequency. Such misclassification would have increased the frequencies of the 4:4 and 2:4:2 patterns relative to the 2:2:2:2 pattern in ditype asci.

The associations between the segregation patterns for mating type and for each of the other eight loci involved in the five crosses studied are shown in Table 3. The data are consistent with random disjunction at the second meiotic division. The percentages of second division segregation are also shown in Table 3, from which the centromere distance for each of the eight loci may be obtained by dividing by two.

The relationship of the ascospores to the apex and base of the ascus was recorded for crosses 1 and 3. The 4:4 and 2:2:2:2 first division segregation patterns for an allele at four different loci segregating in these two crosses, and representing three different linkage groups, are given in Table 4. The apex to base ratios for each pattern do not differ significantly from a 1:1 chance expectation, based on corrected chi-square values. Thus, the data are consistent with random disjunction at the first meiotic division.

TABLE 3

Association of the segregation patterns for the mating type alleles with the segregation patterns for the alleles at eight other loci segregating in 125 concordant asci

Locus and linkage group	Cross	Segregation pattern for the mating type alleles									Percentage* 2nd division segregation
		4:4(31 asci)			2:2:2:2(86 asci)			2:4:2(8 asci)			
		Segregation pattern for the alleles at eight other loci									
		4:4	2:4:2	2:2:2:2	2:2:2:2	4:4	2:4:2	2:4:2	4:4	2:2:2:2	
		Number of asci			Number of asci			Number of asci			
al(102), I	1	1	0	0	6	0	0	1	0	0	0.0
pan(124), IV	3	9	1	0	25	0	1	0	0	0	3.8
	4	11	0	0	27	1	0	4	0	0	..
E(121), VI	1	1	0	0	5	0	1	1	0	0	4.0
	2	5	0	1	13	0	2	2	0	0	..
	3	10	0	0	26	0	0	0	0	0	..
	4	11	0	0	28	0	0	3	1	0	..
	5	3	0	0	11	0	0	1	0	0	..
act(113), V	2	6	0	0	12	2	1	2	0	0	13.0
col(118), V	2	6	0	0	12	3	0	2	0	0	13.0
col(120), IV	3	7	2	1	20	1	5	0	0	0	25.0
col(105), IV	4	4	6	1	16	7	5	2	1	1	48.8
me(123), VI	5	3	0	0	4	4	3	0	0	1	53.3

* Percentage of asci having segregation patterns different than those for mating type.

Discordant asci: There were 84 ditypes and 64 tetratypes among the 148 discordant asci. Positions of discordance (Table 5) occurred as follows: at one end of the ascus, 95 (64.2%); at the center, 17 (11.5%); at both ends, 16 (10.8%); at one end and the center, 15 (10.1%); and at both ends and the center, 5 (3.4%). In the group having discordance at one end, 44 asci were tetratype, whereas in

TABLE 4

Numbers of asci showing various positions of alleles with respect to the apex and base of the ascus for markers segregating at the first division in the concordant asci from crosses 1 and 3

Allele	Cross	First division segregation pattern				Apex to base ratio
		4:4		2:2:2:2		
		Position of allele in the ascus				
		Apex	Base	Apex	Base	
A	1	0	1	6	0	28:15
	3	6	4	16	10	
E(121)	1	1	0	4	1	22:20
	3	6	4	11	15	
col(120)	3	4	3	9	11	13:14
pan(124)	3	5	4	15	10	20:14

TABLE 5

Segregation patterns for mating type initially found in 148 discordant eight-spored asci serially dissected from five different crosses, and reclassification of the 57 least ambiguous discordances to give the most probably correct segregation patterns.

$D =$ ditype; $T =$ tetratype. $m =$ A or a; $+ =$ A or a mating type alleles

Location of discordance	Segregation pattern for mating type initially found in D & T asci†		Number of asci in the five crosses					Total asci	Reclassified segregation pattern for mating type
			1	2	3	4	5		
One end	$\underline{mmmm}\pm\pm++$	D‡	.	.	.	.	.	.	*
		T	0	0	0	3	1	4	4:4
	$\underline{m\pm m\pm}mm++$	D	2	7	3	11	1	24	*
		T	0	1	0	10	7	18	2:2:2:2
	$\underline{m\pm m\pm}++mm$	D	3	2	1	3	1	10	*
		T	0	0	1	0	2	3	2:4:2
	$\underline{m\pm\pm m}++mm$	D	1	5	5	1	5	17	*
		T	0	2	1	11	5	19	2:2:2:2 or 2:4:2
Center	$\underline{mmm\pm m\pm}++$	Dˢ	0	0	1	1	0	2	*
		T	0	0	0	0	0	0	4:4
	$\underline{mm\pm m\pm}m++$	D	2	0	2	2	1	7	*
		T	0	0	2	2	1	5	2:2:2:2
	$\underline{mmm\pm\pm}m++$	D	0	0	1	0	0	1	*
		T	0	0	0	1	0	1	4:4 or 2:2:2:2
	$\underline{mm\pm}+++\underline{mm}$	D‡	.	.	.	.	.	.	*
		T	0	0	0	1	0	1	2:4:2
Both ends	$\underline{m\pm m\pm}m\pm m\pm$	D	0	0	0	1	0	1	*
		T	0	1	0	0	1	2	2:2:2:2
	$\underline{m\pm m\pm}+m\pm m$	D	0	1	0	0	0	1	*
		T	0	0	0	0	0	0	2:4:2
	$\underline{m\pm}+mm++m$	D	0	0	0	2	0	2	*
		T	0	0	0	0	2	2	2:2:2:2 or 2:4:2
	$\underline{m\pm m\pm}+mm+$	D	0	2	2	1	1	6	*
		T	0	0	0	1	1	2	2:2:2:2 or 2:4:2
One end and center	$\underline{mm\pm m\pm}+m+$	Dˢ	2	2	1	2	1	8	*
		T	0	0	0	1	0	1	
	$\underline{mmm\pm\pm}+m$	D	0	0	0	1	0	1	*
		T	0	0	0	0	0	0	
	$\underline{mm\pm m\pm}++m$	D	0	0	0	0	0	0	*
		T	0	0	0	2	1	3	
	$\underline{mm\pm m}++m+$	Dˢ	.	.	.	.	.	.	*
		T	0	0	0	1	0	1	
	$\underline{mmm\pm\pm}++m$	D	0	0	1	0	0	1	*
		T	0	0	0	0	0	0	
Both ends and center	$\underline{mmm\pm m}+\pm+$	Dˢ	.	.	.	.	.	.	*
		T	0	0	0	1	0	1	
	$\underline{m\pm}mm++m+$	D	0	0	1	0	0	1	*
		T	0	0	0	0	0	0	
	$\underline{m\pm}++mm+m$	D	0	1	0	0	0	1	*
		T	0	0	0	1	0	1	
	$\underline{m\pm}++mmm+$	D	0	0	0	1	0	1	*
		T	0	0	0	0	0	0	
Total asci			10	24	22	61	31	148	

* Not reclassified because of ambiguity.
† Type underlined indicates members of a spore-pair (applies only to tetratypes, where spore-pairs are distinguishable).
‡ Scored as concordant 4:4 and 2:4:2, respectively, first division segregation patterns.
ˢ Indistinguishable.

the group having discordance at the center, seven were tetratype. This 44:7 ratio for end and center is significantly different from a 2:1 chance expectation at the 1% level (corrected chi-square = 7.96, 1 df). The total tetratype data for end and center in Table 5 are also significant at the 1% level. Thus, discordance occurred preferentially at an ascus end.

The position of discordance in relation to the apex and base was known in 26 asci with discordances at one end, from crosses 1, 3, and 5. The 17:9 apex:base ratio found is not significantly different from a 1:1 chance expectation (corrected chi-square = 1.88, 1 df). Thus, no preference was shown for one end of the ascus over the other in relation to end discordances.

The frequency of discordant asci was too high to be attributed to dissecting errors. We postulate that the discordance resulted primarily from nuclear passing at the third division and/or from ascospore slippage when the ascospores moved from the biseriate to the uniseriate position (Figure 1). The intermittent occurrence of nuclear passing at the third division should not be unexpected in view of the fact that nuclear passing at both the second and third divisions occurs regularly in the wild type ascus of this species.

The 57 tetratype asci with discordance at one end, or at the center, or at both ends were reclassified by assuming that these discordances were the direct result of nuclear passing at the third division and/or of ascospore slippage; the reclassified ascospore patterns obtained by making the simplest rearrangements under this assumption are shown, with respect to the mating type alleles, in Table 5. Reclassification was not done for any of the 84 ditypes, nor for the remaining seven tetratypes having discordances at one end and the center or at both ends and the center because of the ambiguity encountered. After reclassification, 4 or 5 of the 57 tetratype asci had the 4:4 segregation pattern for the mating type alleles, and 52 or 53 had the 2:2:2:2 or 2:4:2 patterns. It will be recalled (Table 2) that in the concordant tetratype asci there were eleven 4:4 segregation patterns for the mating type alleles and thirty-five 2:2:2:2 or 2:4:2 patterns. A contingency table used to test for independence of first division segregation patterns and ascus types (reclassified tetratypes and concordant tetratypes) showed a significant chi-square value (4.3, 1 df) at the 5% level. Thus discordance occurred significantly less frequently in asci showing the 4:4 pattern for the mating type alleles than in asci showing the other two patterns.

It seems likely that the probability of discordance would be a function of the proximity of the nuclei during ascus development. Thus, relatively little discordance was found in the 4:4 first division segregation pattern believed to result from the even dispersal of the four telophase nuclei at the second division (Figure 1). On the other hand, the 2:2:2:2 and 2:4:2 patterns showed discordances frequently at the ends of the asci, where nuclei are closely associated (Figure 1), but less frequently at the center, where nuclei are not so close together.

Genetic estimates of spindle behavior frequencies: In order to base such estimates upon both concordant and discordant asci, only tetratypes were used, because of the difficulty of identifying sister ascospores in discordant ditypes. There were 46 concordant tetratypes (Table 2) and 57 reclassifiable discordant tetratypes, 33 of which were reclassified into the most probably correct segregation pattern and 24 of which could only be reclassified as either of two most probably correct patterns (Table 5). The frequency of the 4:4 pattern could be determined rather precisely, since either 15 or 16 of the 103 tetratypes under consideration (14.6% or 15.5%) showed this pattern. The frequencies of the other two patterns could not be precisely determined because of the dichotomy in reclassification of 24 of the asci; these frequencies are therefore stated only as ranges, namely, 54.4% to 77.7% for the 2:2:2:2 pattern and 7.8% to 30.1% for the 2:4:2 pattern, depending upon the pattern to which the 24 asci are assigned.

The values 14.6% or 15.5% for the 4:4 pattern represent our genetic estimate of the frequency of tandemly oriented second division spindles, which we have postulated give rise to the 4:4 pattern; these values are in reasonably good agreement with Singleton's cytological determination of "about 10%" tandemly oriented spindles (Braver 1952). Overlapping spindles apparently occurred much more frequently in ascus development, during which the less extreme type of nuclear passing predominated (54.4% to 77.7%); this type of passing is believed to result in the 2:2:2:2 pattern (Figure 1, middle row).

Four-spored asci: Crosses heterozygous for E (121) produce some four-spored asci in the same perithecia in which the generally more numerous eight-spored asci are produced. 54 and 49 four-spored asci, respectively, were analyzed from crosses 4 and 5, both of which were heterozygous for E. These asci showed the same genetic characteristics recently found to be typical of four-spored asci from crosses in which E is not segregating (Calhoun and Howe 1968). The four-spored asci from both of these origins, as well as eight-spored asci, have all given consistently similar centromere distances. Thus, four-spored asci have not been shown to be detectably different whether originating from crosses in which E is segregating or from crosses lacking E. The basis for this apparently frequent variation in expression of the E allele is not yet clear.

Significance of gene E: The major significance of gene E probably lies in its pronounced effect upon meiosis and certain other events during ascus development. The dimensions of the ascus become modified. Nuclear passing at the third division, and sometimes at the second division, is apparently eliminated, and the nuclei are subsequently delimited singly instead of pair-wise during formation of the ascospore walls. The homokaryotic ascospores thereby produced have the practical advantage of being much more amenable to genetic analysis than are the heterokaryotic ascospores of the four-spored ascus.

The technical assistance of Mrs. J. A. Hancock is gratefully acknowledged. Mr. Calhoun was the recipient of a National Science Foundation Traineeship, Grant GZ-45.

SUMMARY

Crosses heterozygous for the gene E produce eight-spored asci having two especially noteworthy characteristics: (1) both first and second division segregations are manifested by all three segregation patterns, 4:4, 2:2:2:2 and 2:4:2 and (2) about half of the asci are discordant, i.e., the two members of at least one ascospore-pair (sister ascospores) are nonadjacent. The 4:4 pattern arises from tandemly oriented spindles at the second meiotic division which have a genetically determined frequency of about 15%. Much more commonly, however, spindles overlap at the second division and give rise to the two types of nuclear passing that produce the 2:4:2 and 2:2:2:2 patterns, the latter more frequently. Discordance probably results from nuclear passing occurring intermittently at the third division and/or from ascospore slippage. Because of the reliability of the mating type locus as a centromere marker, discordant tetratype asci may usually be reclassified to fit the most probably correct ascospore arrangement. The four-spored asci which are usually also present in crosses heterozygous for E do not differ detectably in genetic behavior from four-spored asci from crosses in which E is not segregating.

LITERATURE CITED

BERG, C. M., 1966 Biased distribution and polarized segregation in asci of *Sordaria brevicollis*. Genetics **53**: 117–129.

BRAVER, N. B., 1952 Genetic segregations in *Neurospora tetrasperma*. M.S. Thesis, University of Missouri, Columbia, Missouri.

CALHOUN, F., 1968 Nuclear movement during development of four-spored and eight-spored asci in *Neurospora tetrasperma*. M.S. Thesis, University of Georgia, Athens, Georgia.

CALHOUN, F., and H. B. HOWE, 1968 Ascospore linearity in the four-spored ascus of *Neurospora tetrasperma*. Genetica **39**: 245–249.

DODGE, B. O., 1939 A new dominant lethal in Neurospora. The E locus in *N. tetrasperma*. J. Heredity **30**: 467–474.

DODGE, B. O., J. R. SINGLETON, and A. ROLNICK, 1950 Studies on lethal E gene in *Neurospora tetrasperma*, including chromosome counts also in races of *N. sitophila*. Proc. Am. Phil. Soc. **94**: 38–52.

HAWTHORNE, D. C., and R. K. MORTIMER, 1960 Chromosome mapping in Saccharomyces: centromere-linked genes. Genetics **45**: 1085–1110.

HOWE, H. B., and P. HAYSMAN, 1966 Linkage group establishment in *Neurospora tetrasperma* by interspecific hybridization with *N. crassa*. Genetics **54**: 293–302.

THE EFFECT OF THE CURLY INVERSIONS ON MEIOSIS IN DROSOPHILA MELANOGASTER

I. INTRACHROMOSOMAL EFFECTS ON RECOMBINATION

By CLAES RAMEL

INTRODUCTION

INTERCHROMOSOMAL effects of inversions in *Drosophila melanogaster* have been analyzed in a series of investigations by the present author (RAMEL, 1962 a, 1962 b, 1965, 1966; RAMEL and VALENTIN, 1965). For many of these studies the second chromosome inversions (2 L+2 R) Curly have been used. A detailed analysis of one and the same inversion chromosome has had definite advantages. In particular it has been of great value, if not a necessity, for the interpretation of different kinds of interchromosomal effects in relation to each other. In view of this fact, it seemed desirable to extend the analyses of the Curly inversions from some new angles. The present investigations can therefore be considered as a continuation of the work reported previously, although more strictly concentrated to the analysis of the Curly inversions.

The two inversions constituting Ins (2 L+2 R) Cy can be separated by crossing over. These inversions, In (2 L) Cy and In (2 R) Cy, were used in the present investigation to study their effect on crossing over across the centromere that is, the *intra*chromosomal effect of the inversions on crossing over. The problem dealt with in this investigation is as follows.

In conformity with the well established interchromosomal effects of heterozygous inversions in general, the Curly inversions cause an increased crossing over and a decreased interference in heterologous chromosomes. It has been emphasized by the present author that the experimental data concerning this interchromosomal effect on crossing over, cannot be interpreted exclusively on the basis of a "mechanical" interference with the meiotic chromosome pairing, such as non homologous associations. Although such disturbances no doubt take place with heterozygous chromosomal aberrations, like inversions (cf. RAMEL, 1962 b), the experimental evidence rather indicates that the ultimate effect of the inversions depends on some physiological process. It was proposed (RAMEL, 1965) that this physiological interaction between heterologous chromosomes reflects a competition for an agent essential in the process of crossing over. With reference to the crossing over hypothesis presented by UHL (1965) it was furthermore suggested that this agent might be protein links between DNA helices, attached end to end in tandem. It was postulated that an altered competition for these links was created primarily as a result of the delayed synapsis of heterozygous aberrations, particularly inversions. Such a kind of competition for a common agent, involved in recombination, would also explain that the chiasma frequency often tends to remain constant between cells with a compensation mechanism operating between bivalents (MATHER, 1936; MATHER and LAMM, 1935).

If this effect of heterozygous inversions rests upon such a mechanism, there is no reason why it should be restricted to an influence on heterologous chromosomes. Similar effects should operate also within the same chromosome. The long autosomes are particularly suitable for a study of this problem, as the effects of a heterozygous inversion in one arm can be analyzed across the centromere in the opposite arm. It can be assumed that the non inverted arm would be less influenced mechanically by the actual pairing procedure of the inverted section than if crossing over was measured outside an inversion in the same chromosome arm.

Material and methods

The inversion stocks used for the present purpose, In (2 L) Cy and In (2 R) Cy, were identical with the ones used in a previous series of experiment (RAMEL and VALENTIN, 1965). They were obtained through crossing over from one female containing both inversions in tandem, Ins (2 L+2 R) Cy, al^2 lt^s sp^2. (For further details, see RAMEL and VALENTIN, 1965). The left Curly inversion, In (2 L) Cy, has its left break point between S and dp and its right break point left of b. The right inversion, (2 R) Cy is larger, including most of the right arm.

Females homozygous for al (0.0), dp (13.0), b (48.5), pr (54.4), c (75.5), px (100.5), sp (107.0) were mated to males heterozygous for In (2 R) Cy. Crossing over was subsequently measured for the intervals al-dp-b-pr at the presence or absence of (2 R) Cy. A corresponding study of the intervals pr-c-px-sp at the presence or absence of (2 L) Cy turned out to be less suitable because of the difficulty of identifying curved (c) at the presence of the Curly phenotype. Therefore another marker stock carrying b (48.5), cn (57.5), vg (67.0), bw^D (104.5) was used in a similar way for the investigation of crossing over in the right arm at the presence or absence of (2 L) Cy.

The females of the experimental crosses were aged for 6 days and subsequently mated in 1/3 litre bottles with 15 females and 15 males per bottle. The flies were transferred to new bottles for four egg laying periods of two days each. The experiments were performed at $25° \pm 1°$.

RESULTS

The results of the experiments are shown in tables 1 and 2. It can be seen in table 1 that heterozygosity for the left inversion causes an increase of crossing over in the right arm. This increase involves the interval cn-vg and reaches significance for females of the age 8—10 and 10—12 days as well as for the total sum ($P < 0.01$). For the interval vg-bw^D, all the age groups, as well as the total sum, show a significantly higher crossing over frequency with the heterozygous inversion as compared to the control.

The interval b-cn is partially located in the same arm as the left Cy inversion, although fully outside the inverted segment. In spite of this the crossing over frequency is very much reduced by the presence of the inversion (see Table 1). It is obvious that the effect of the inverted segment on meiotic synapsis reaches far outside the limits of the proper inversion. Although only about a third of the interval b-cn extends to the right arm, the crossing over frequency seems to be too low, even to account for that part of the interval, located in the right arm. The crossing over value in the series with (2 L) Cy is, in all cases, considerably lower than a third of the control value. This would indicate that synapsis is affected also across the centromere from the left to the right arm. This fact does not point to a synaptic process, invariably proceeding from the centromere towards the ends of the chromosome arms, as has been suggested by some authors. Considering the experiment with In (2 R) Cy from this point of view, it should be pointed out that no

TABLE 1. *Analysis of crossing over and coincidence with* $\frac{b\ cn\ vg\ bw^D}{+\ +\ +\ +}$ ♀♀ *with and without In (2 L) Cy, crossed to* $\frac{b\ cn\ vg}{b\ cn\ vg}$ ♂♂.

Age of females	Per cent crossing-over								Coincidence cn−vg−bwD		Number counted	
	b−cn		cn−vg		vy−bwD							
	+	Cy (2 L)	+	Cy (2 L)	+	Cy (2 L)			+	Cy (2 L)	+	Cy (2 L)
6—8	8.6±0.8	1.8±0.4	11.3±0.9	11.6±1.0	36.1±1.5	40.4±1.5			0.39±0.08	0.65±0.09	1340	1028
8—10	4.9±0.3	1.0±0.2	8.3±0.5	10.1±0.5	33.4±0.8	37.7±0.8			0.43±0.05	0.47±0.05	4261	3658
10—12	4.0±0.3	0.5±0.1	8.5±0.5	10.4±0.5	32.5±0.8	36.4±0.8			0.50±0.07	0.53±0.06	3289	3208
12—14	2.6±0.3	0.1±0.1	7.5±0.7	7.0±0.6	29.5±1.2	37.2±1.2			0.50±0.09	0.44±0.09	2263	1564
Total	4.6±0.2	0.8±0.1	8.6±0.3	9.8±0.3	32.7±0.5	37.5±0.5			0.46±0.03	0.51±0.03	11153	9458

TABLE 2. Analysis of crossing over and coincidence with $\frac{al\ dp\ b\ pr\ c\ px\ sp}{+\ +\ +\ +\ +\ +\ +}$ ♀♀, with and without In (2R) Cy, crossed to $\frac{al\ dp\ b\ pr\ c\ px\ sp}{al\ dp\ b\ pr\ c\ px\ sp}$ ♂♂.

Age of females	Per cent crossing-over						Coincidence				Number counted	
	al–dp		dp–b		b–pr		al–dp–b		dp–b–pr			
	+	Cy (2 R)	+	Cy (2 R)	+	Cy (2 R)	+	Cy (2 R)	+	Cy (2 R)	+	Cy(2R)
6—8	11.6±1.6	10.4±1.6	26.8±2.2	33.0±2.5	6.9±1.2	3.1±0.9	0.32±0.14	0.66±0.21	0.51±0.23	0.83±0.40	421	355
8—10	14.4±1.4	13.2±1.5	27.8±1.8	30.0±2.0	5.4±0.9	4.4±0.9	0.16±0.08	0.29±0.11	0.31±0.17	1.02±0.31	647	523
10—12	15.2±1.4	14.6±1.5	25.7±1.7	34.7±2.0	3.6±0.7	4.6±0.9	0.07±0.05	0.25±0.09	0.24±0.23	0.57±0.22	686	548
12—14	9.8±1.5	10.4±1.4	28.5±2.3	34.5±2.1	3.3±0.9	5.0±1.0	—	0.22±0.11	0.53±0.34	0.93±0.26	400	498
Total	13.2±0.7	12.4±0.8	27.1±1.0	33.1±1.1	4.7±0.5	4.4±0.5	0.10±0.04	0.32±0.06	0.36±0.11	0.82±0.14	2154	1924

corresponding decrease of crossing over in the vicinity of the centromere in the opposite arm was noticeable in this case.

The material from the experiment with In (2 R) Cy is rather small, but the trend is the same as in the case above with (2 L) Cy. The results in Table 2 show that the presence of (2 R) Cy causes an increase of crossing over for the interval dp-b. This increase is highly significant for the total sum ($P < 0.001$). The material does not show any heterogeneity for crossing over in this interval, caused by the age of the females. No difference between the inversion series and the control can be established for the other two intervals, al-dp and b-pr.

In the experiment with the left inversion (Table 1) there is no clear effect on interference. Only for females of the age 6—8 days is there a significantly higher coincidence value in the inversion series as compared to the control ($0.05 > P > 0.02$).

In Table 2 it can be seen that the concidence values are higher for the series with (2 R) Cy throughout the material. For the total sum, independent of the age of the females, the differences are significant at the 1 per cent level both for al-dp-b and for dp-b-pr.

DISCUSSION

The experimental data indicate that heterozygous paracentric inversions have an intrachromosomal effect across the centromere. This effect is of a similar nature as the interchromosomal effect of inversions on recombination that is, crossing over is increased and interference weakened. These data are supported by some previous investigations. Thus KIKKAWA (1934) has pointed out the occurrence of a compensation mechanism for crossing over within the same chromosome arm. If recombination was increased at one end of a chromosome by means of X irradiation or temperature changes, it tended to decrease simultaneously at the other end of the same chromosome arm. Likewise KOMAI and TAKAKU (1942) noted an increased crossing over in the distal part of the long X chromosome in *Drosophila virilis* at the presence of heterozygous inversions proximally.

DOBZHANSKY and STURTEVANT (1931) and GLASS (1933) have reported interchromosomal effects on crossing over in translocations, which may be of relevance in this connection. The introduction of a heterozygous inversion into the non translocated chromosome 2 in a 2; 3-translocation, caused an increased crossing over in chromosome 3. Thus an effect across the centromeres in the translocation configuration was brought about by the inversion. The situation is, however, somewhat more complicated than in the present case, where only inversions are involved.

It may be concluded from the experimental data that crossing over relationships are present within the chromosomes that bear a close resemblance to interchromosomal effects on crossing over. The similarity between the intrachromosomal effects of inversions dealt with in this report, and the corresponding effects on heterologous chromosomes, makes it highly probable that the same mechanism is involved in both cases. Consideration must be taken to this fact when interpreting interchromosomal effects on recombination.

As pointed out in the introduction, the interchromosomal effects of inversions on crossing over are difficult to explain solely as a direct result of mechanical interactions during meiotic synapsis. The conformity of the effects, exerted by inversions between as compared to within chromosomes render further support for this view. On the other hand the results on intrachromosomal effects are in accordance with expectation if physiological interactions of the kind outlined previously (RAMEL, 1965) are involved.

Acknowledgements. — The author is greatly indepted to Mr. T. LINDBERG for his assistance in the experiments. This work has been supported by a research grant from the Swedish Natural Science Research Council.

SUMMARY

Investigations were made of the effect of the separate Curly inversions, In (2 L) Cy and In (2 R) Cy, on crossing over in the opposite arm of chromosome 2, across the centromere. It was found that the intrachromosomal effect on crossing over exerted by the inversions, was in accordance with the corresponding effect on crossing over in heterologous chromosomes, studied previously. Thus an increased crossing over was obtained at the presence of the inversions and the results also indicated a decreased interference.

Literature cited

DOBZHANSKY, T. and STURTEVANT, A. H. 1931. Contributions to the genetics of certain chromosome anomalies in *Drosophila melanogaster*. II. Translocations between the second and third chromosomes of *Drosophila* and their bearing on *Oenothera* problems. — Publ. Carneg. Inst. 421: 29—59.

GLASS, H. B. 1933. A study of dominant mosaic eye-colour mutants in *Drosophila melanogaster*. II. Tests involving crossing-over and non-disjunction. — J. Genet. 28: 69—112.

KIKKAWA, H. 1934. Studies on non-inherited variation in crossing-over in *Drosophila*. — Ibid. 28: 329—348.

KOMAI, T. and TAKAKU, T. 1942. On the effect of the X-chromosome inversions on crossing-over in *Drosophila virilis*. — Cytologia 12: 357—365.

MATHER, K. 1936. Competition between bivalents during chiasma formation. — Proc. Roy. Soc. (London) Ser. B, 120: 208—227.

MATHER, K. and LAMM, R. 1935. The negative correlation of chiasma frequencies. — Hereditas 20: 65—70.

RAMEL, C. 1962 a. Interchromosomal effects of inversions in *Drosophila melanogaster*. I. Crossing-over. — Ibid. 48: 1—58.

— 1962 b. Interchromosomal effects of inversions in *Drosophila melanogaster*. II. Non-homologous pairing and segregation. — Ibid. 48: 59—82.

— 1965. The mechanism of interchromosomal effects on crossing-over. — Ibid. 54: 237—248.

— 1966. The interchromosomal effect of inversions on crossing-over in relation to non homologous pairing in *Drosophila melanogaster*. — Ibid. 54: 293—306.

RAMEL, C. and VALENTIN, J. 1965. Interchromosomal effects of the Curly inversions on crossing-over in *Drosophila melanogaster*. — Ibid. 54: 307—313.

UHL, C. H. 1965. Chromosome structure and crossing-over. — Genetics 51: 191—207.

Etiology of nondisjunction: lack of evidence for genetic control[1]

R. C. JUBERG and LOUISE M. DAVIS

Introduction

Several lines of evidence suggest that the occurrence of meiotic nondisjunction in man is not always a random event (DAY, 1966). Observations of more than one aneuploid member in either a sibship or a kindred may indicate the existence of genetic control of chromosomal segregation (THERMAN et al., 1961; HECHT et al., 1964).

The question of the presence of genetic control of meiotic segregation in a panmictic population could possibly be answered in several different ways. First, subjects who have resulted from meiotic nondisjunction would be identified, i.e., those aneuploid persons without mosaicism. Then the next step would be the computation of the extent of consanguinity in their ancestors. However, in the general

---

[1] Supported in part by a training grant from the Children's Bureau, Department of Health, Education and Welfare, to the Department of Pediatrics, University of Michigan (R.C.J.).

population this approach would be handicapped seriously by the inability of the average couple to provide a sufficiently detailed family history, with the result that all consanguineous marriages would not be identified. In addition, some couples might be reluctant to admit their knowledge of such relationships. Furthermore, suitable comparisons might be difficult to obtain. In an English series of more than 600 mongols, PENROSE (1961) found that the parents of three were related as cousins, and five of the paternal grandparents and ten of the maternal grandparents were cousins, although the degrees of cousin relationship were not stated. A comparison group was not mentioned. In a Swedish population FORSSMAN and ÅKESSON (1967) found no more consanguineous matings, specifically of first cousins or nearer relationship, among the parents of mongols than they estimated were present in the general population.

These disadvantages would be avoided if the aneuploid propositi were ascertained in families with extensive and accurate ancestral records. The advantage of identification of aneuploids with a similar karyotype would also be possible in a large population. In the event that all members of the population lived in a similar environment, then an ideal comparison group would be available.

Second, a population with a significant number of consanguineous matings might be surveyed for the frequency of aneuploids. SCHULL and NEEL (1965) extensively studied a Japanese population but provided no specific evidence on this question. In another study of a Japanese population, MATSUNAGA (1966) found fewer parents who were first cousins among the younger mothers of mongols than he estimated would be expected if recessive genes contributed to the etiology of nondisjunction.

Third, an isolate either known or alleged to contain a significant number of consanguineous matings might be used in order to calculate the frequency of birth of aneuploid progeny. This approach would be satisfactory only if (1) the reproductive practices were shown to be similar to those in the general population or in the population used for comparison and (2) the precise number of births per unit of time were known. Finally, without a fairly reliable estimate of the extent of consanguinity, the data would be difficult to interpret. KWITEROVICH *et al.* (1966) used this method in an isolate and found the frequency of mongolism similar to that usually reported for the general population.

The purpose of this report is to present the results of an investigation using the first approach. A population with voluminous and reliable family records was first identified. Then the most readily diagnosable phenotype produced by aneuploidy and compatible with long survival, the mongol, was selected for ascertainment. The population size of approximately 5500 persons also allowed a suitable comparison group to be drawn from the same isolate.

Materials and methods

At the beginning of this study there were 30 Old Order Amish church districts in Elkhart and Lagrange Counties, Indiana (DER NEUE AMERIKANISCHE CALENDER AUF DAS JAHR UNSERES HEILANDES JESU CHRIST, 1964). Initially, we took a census of the population in each church district by personal interviews with reliable informants, whom we knew from previous investigations. We recorded for each couple the surname, maiden name, given names, approximate age by decade from 20–29 to 80–89 years of the adults, and the number and sex of all livebirths. We then sent each roster to the district bishop for verification. All were returned, and the only corrections were of the few deaths, births, and marriages which had occurred in the interval since the original interview. In addition, during each visit we described the phenotypic manifestations of mongolism, and we asked the family for the name of anyone fitting the description.

Sixteen persons with mongolism were ascertained. We were confident that all living mongols were identified since the Amish rarely institutionalize abnormal offspring. We were well acquainted with informative members in each of the church districts, and each Amish family is well acquainted with the other families of the church district. However, physical examination of all persons in the population was not possible.

A specimen of blood was obtained from each of the mongols, and cultures were established by a modification of the method described by MOORHEAD et al. (1960).

A comparison family was selected for each of the 16 families with a mongol by matching for sibship size and age of each mother at the time of the study. For example, for the first patient listed in Table I, subject D.W.H., who was a member of a sibship of 14 and whose mother was 64 years old at the time of the study, a comparison family was selected at random from the families of sibship size of 14 in which the mother was 60–69 years old at the time of the study. Most of the comparison families were unfamiliar to us, but when they were approached, told about the study, and asked to provide as much family information as possible, they were cooperative. In the few instances where the census data were erroneous regarding either the number of progeny or maternal age, another family was selected from the listings. The availability of appropriate comparisons and their cooperation illustrate the advantages of the Amish population for this type of investigation.

Then we interviewed each of the 32 families. They provided information identifying all ancestors, such as name, date of birth, date of death, and marital partner, from their knowledge and their own family history books. The Amish have a particular interest in genealogy, and they manifest this by the publication of the record of the descendants of a certain pair of ancestors. They do this so diligently that although certain persons will naturally occur in several different publications, each compiled by different descendants, we found remarkably few inconsistencies during the investigation. Some of the older books contain records of ancestors who were born in the 1700's and immigrated to the United States from Switzerland. We used more than 50 of these books, with publication dates ranging from 1900 to 1964, to complete the ancestries beyond those furnished by each couple. Thus, by the use of these records we constructed the ancestry for all of the couples. Ancestors were identified for as many as 8 generations in some cases and 10 generations in a few. Of course, we knew whether the family contained a mongol or a comparison because of our familiarity with the families of the mongols. The inbreeding coefficients were calculated manually.

Results

There were 11 male and 5 female mongols (Table I). The expected effect of maternal age and birth order was shown by the test described by HALDANE and SMITH (1948) (sum of birth ranks of affected siblings = 462, theoretical mean sum of birth ranks of affected siblings = 324, standard deviation of theoretical mean = 46). At the time of the study, 10 of the mothers of the mongols were more than 45 years old.

Range in age of the mothers at the time of conception of the propositi was 21 to $43\frac{3}{4}$ years, and the mean age at conception was $35\frac{1}{2}$ years. Range in age of the fathers at the time of conception of the mongols was $25\frac{1}{2}$ to 47 years, and the mean age at conception was $39\frac{1}{2}$ years.

Cytogenetic evaluation showed a modal count of 47 chromosomes, with an extra G chromosome in each subject. All nonmodal counts appeared to be the result of either random loss or gain, and there was no evidence of mosaicism in any of the propositi.

A tabulation of ancestors who were positively identified confirmed the assumption that the ancestors of the comparison group could be traced as thoroughly as those of the mongols. The numbers in Table II refer only to identification, and, therefore, an ancestor who was common to two or more individuals was counted that number of times. The table shows our success in constructing the pedigrees. The reduction from the theoretical value means that we could not find some ancestors in any of the family records, and the diminution does not reflect consanguinity. As a matter of fact, we were slightly more successful in locating the ancestors of the comparison group than we were in locating those of the mongols.

Table I. Sixteen patients with the clinical diagnosis of mongolism ascertained in an Old Order Amish isolate in Elkhart and Lagrange Counties, Indiana.

Patient	Sex	Age (years)	Sibship position	Sibship size	Parental age at conception	
					Mother (years)	Father (years)
D.W.H.	M	1/4	14	14	43 3/4	43
M.E.B.	M	2	3	3	31 1/2	28 3/4
P.S.H.	M	4 1/2	2	2	40 1/4	39 3/4
D.E.M.	M	9	8	8	43	42 3/4
K.K.	M	10 1/2	2	3	21	25 1/2
O.A.M.	M	14 1/4	2	7	32 1/2	33 3/4
E.M.B.	M	15	7	7	42 1/4	40 1/4
M.O.W.	M	15	4	4	42 1/4	46
D.M.	M	16 1/4	4	6	33 1/2	39 3/4
Q.L.H.	M	16 1/4	5	6	30 3/4	29 3/4
F.W.M.	M	18 1/2	5	5	37 1/4	40 3/4
A.R.B.	F	2 1/2	1	1	29 1/2	29 1/2
K.M.M.	F	7	3	6	23 1/4	25 3/4
R.L.P.	F	17 1/4	9	10	42 3/4	44
A.M.	F	18 1/4	6	8	35	38 1/4
M.S.	F	26 1/2	3	4	41	47
Mean	M(11) F(5)				35 1/2	39 1/2

Table II. The number of ancestors of 16 mongol propositi and of 16 comparison sibships identified through published Amish family records.

Relationship	Expected No.	Propositi		Comparisons	
		No.	Percentage of theoretical value	No.	Percentage of theoretical value
Parents	32	32	100	32	100
Grandparents	64	62	97	64	100
Great-grandparents	128	122	95	128	100
Great-great-grandparents	256	207	81	243	95
Great-great-great-grandparents	512	294	57	371	72

Table III. Coefficients of inbreeding for the preceding three generations of 16 mongol propositi and 16 comparisons.

Relationship	Mongol pedigrees No. in which any inbreeding	f (mean ± SD)	Comparison pedigrees No. in which any inbreeding	f (mean ± SD)	Difference in estimate of f
Mother	8/16	0.0050 ± 0.0039	13/16	0.0051 ± 0.0019	−0.0000
Father	9/16	0.0036 ± 0.0020	14/16	0.0059 ± 0.0017	−0.0021
Maternal grandmother	3/16	0.0008 ± 0.0005	4/16	0.0009 ± 0.0004	−0.0000
Maternal grandfather	3/16	0.0024 ± 0.0014	10/16	0.0038 ± 0.0012	−0.0014
Paternal grandmother	6/16	0.0040 ± 0.0017	7/16	0.0070 ± 0.0024	−0.0030
Paternal grandfather	6/16	0.0036 ± 0.0021	8/16	0.0016 ± 0.0006	+0.0020
Maternal maternal great-grandmother	1/16	0.0002 ± 0.0002	3/16	0.0009 ± 0.0006	−0.0000
Maternal paternal great-grandmother	0/16	0	1/16	0.0001 ± 0.0000	−0.0000
Maternal maternal great-grandfather	1/16	0.0010 ± 0.0010	4/16	0.0026 ± 0.0020	−0.0016
Maternal paternal great-grandfather	1/16	0.0005 ± 0.0005	0/16	0	+0.0000
Paternal maternal great-grandmother	2/16	0.0003 ± 0.0003	4/16	0.0026 ± 0.0014	−0.0022
Paternal paternal great-grandmother	1/16	0.0001 ± 0.0005	3/16	0.0017 ± 0.0011	−0.0016
Paternal maternal great-grandfather	3/16	0.0007 ± 0.0005	5/16	0.0042 ± 0.0022	−0.0035
Paternal paternal great-grandfather	3/16	0.0010 ± 0.0007	1/16	0.0005 ± 0.0005	+0.0000
Sibship	11/16	0.0038 ± 0.0015	15/16	0.0066 ± 0.0015	−0.0028

Table III shows the coefficients of inbreeding (f) for the ancestors in three generations preceding the mongols and their comparisons. The largest mean inbreeding coefficient was for the paternal grandmothers in the comparison group. The value of 0.0070 is less than that of a two and one-half cousin relationship. In the pedigrees of the mongols the mean f of the parents (0.0043) was greater than that of the grandparents (0.0027), which, in turn, was greater than that of the great-grandparents (0.0005). Similarly, in the pedigrees of the comparisons the mean f of the parents (0.0055) was greater than that of the grandparents (0.0033), which, in turn, was greater than that of the great-grandparents (0.0016). This result was anticipated because of the increasing difficulty in establishing the identity of ancestors in the more remote generations. Hence, we do not believe that these results indicate that there is necessarily more inbreeding in the current generations than there was in the past.

A significant difference in mean f was not found between comparable ancestors in the pedigrees of the mongols and of the comparisons for any of the 14 ancestors tested. However, the mean f of the ancestor of the mongol sibship was greater than that of the ancestor of the comparison sibship in only three instances, a finding contrary to expectation if inbreeding contributes to the etiology of meiotic nondisjunction. The distribution of these values of f is significant by the sign test (SIEGEL, 1956).

The inbreeding coefficients for the sibships of the mongols and their comparisons are similar to those of the parental generation, but, of course, the former are not relevant to the etiology of meiotic nondisjunction in the parental germ cells.

Two of the mongols were related to each other as second cousins, and another two were similarly related, leaving 12 sibships not shown to be related. Among the comparison group, in three instances there were two sibships related as second cousins, leaving 10 sibships not shown to be related.

Discussion

Since the advent of human cytogenetics, numerous families and kindreds have been reported with more than one aneuploid member, excluding the families in which translocation predisposes to abnormal segregation. While a complete tabulation of these is beyond the scope of the present report, no instance of consanguinity has come to our attention. Whether the number of these is really more than would be expected has not been determined, if we assume that from 0.4% to 1% of livebirths have a numerical chromosomal abnormality (COURT BROWN, 1967, pp. 1–31; SERGOVICH et al., 1969). Systematic epidemiological studies have not yet been reported in which the population at risk, which constitutes the denominator of the frequency fraction, has been determined.

The incidence of mongolism among births in the Amish isolate was not determined since such an estimate would have required a different approach and the accumulation of different data, including the enumeration of all births during a defined period of time, a reliable description of all progeny who died during the interval, and some estimate of the gains and losses from migration. KWITEROVICH et al. (1964) ascertained 16 mongols in an Amish isolate of Holmes County, Ohio, among the population estimated at 10,000.

The average inbreeding coefficient of 0.0195 for an entire population of Amish in Adams County, Indiana, reported by JACKSON et al. (1968) is larger than that of any of the ancestors in the present study. Their data were based on analysis of 276 matings, while 480 matings were evaluated in the present study. Although the two groups of Amish live close to each other, perhaps their mating practices differ considerably. At least the relative frequency of their surnames is different.

The extent to which the mongols themselves were related to each other compared with the extent to which the subjects in the comparison group were related was determined because of the possibility that one of the founders of the isolate possessed a gene controlling disjunction which was eventually transmitted selectively to one or both parents of the mongols. However, approximately as many comparison subjects were related to each other as were the propositi. In addition, no ancestors unique to the mongols were found.

The negative results of this study certainly do not mean that genetic control of chromosomal segregation does not exist in man. Since genetic regulation of meiotic behavior has been identified in other organisms (STURTEVANT, 1936; CATCHESIDE, 1944; RHOADES, 1952; GRELL, 1959; SANDLER et al., 1968), it is a reasonable assumption that it also exists in man. Nevertheless, three conclusions seem warranted from the present data.

The first is that a genetic mechanism controlling the abnormal segregation of the G chromosome causing mongolism is not present in the sample of the population studied. If the effect of such a gene were not specific for either a particular type or size of chromosome, then the statement could be made more general to include all chromosomal segregation. Second, such a gene or genes may not be very common in the population based upon our sample of 16 sibships, which we know generally to be independent of each other, and therefore derived from a large number of ancestors. Third, since English, Swedish, Japanese, and now North American populations have been similarly and independently investigated with negative results, the frequency of recessive genes controlling chromosomal segregation at meiosis must be relatively rare, if, in fact, they do exist.

Acknowledgements

The authors express their appreciation to Dr. MARGERY W. SHAW for commenting upon the data, to Dr. F. CLARKE FRASER for reviewing the manuscript, to Dr. PAUL H. MARTIN for supporting part of the study, to Mrs. MERRILY G. HART for technical assistance, and to the numerous Amish and Mennonite families for the use of their family records.

References

CATCHESIDE, D. G.: Polarized segregation in an ascomycete. Ann. Bot. *8:* 119–130 (1944).
COURT BROWN, W. M.: Human population cytogenetics (North-Holland Publishing Company, Amsterdam 1967).
DAY, R. W.: Symposium: etiology of chromosomal abnormalities. The epidemiology of chromosome aberrations. Amer. J. hum. Genet. *18:* 70–80 (1966).
DER NEUE AMERIKANISCHE CALENDER AUF DAS JAHR UNSERES HEILANDES JESU CHRIST (J. A. Raber, Baltic, Ohio 1964).
FORSSMAN, H. and ÅKESSON, H. O.: Consanguineous marriages and mongolism, pp. 23–34. *In* G. E. W. WOLSTENHOLME and R. PORTER, eds.: Mongolism, Ciba Foundation Study Group No. 25 (Little, Brown, Boston 1967).
GRELL, R. F.: Nonrandom assortment of non-homologous chromosomes in *Drosophila melanogaster*. Genetics 44: 421–435 (1959).
HALDANE, J. B. S. and SMITH, C. A. B.: A simple exact test for birth-order effect. Ann. Eugen. *14:* 117–124 (1948).
HECHT, F.; BRYANT, J. S.; GRUBER, D. and TOWNES, P. L.: The nonrandomness of chromosomal abnormalities. Association of trisomy 18 and Down's syndrome. New Engl. J. Med. *271:* 1081–1086 (1964).
JACKSON, C. E.; SYMON, W. E.; PRUDEN, E. L.; KAEHR, I. M. and MANN, J. D.: Consanguinity and blood group distribution in an Amish isolate. Amer. J. hum. Genet. *20:* 522–527 (1968).
KWITEROVICH, P. O., JR.; CROSS, H. E. and MCKUSICK, V. A.: Mongolism in an inbred population. Bull. Johns Hopkins Hosp. *119:* 268–275 (1966).
MATSUNAGA, E.: Down's syndrome and maternal inbreeding. Acta genet. med. gemellog. *15:* 224–229 (1966).
MOORHEAD, P. S.; NOWELL, P. C.; MELLMAN, W. J.; BATTIPS, D. M. and HUNGERFORD, D. A.: Chromosome preparations of leukocytes cultured from human peripheral blood. Exp. Cell Res. *20:* 613–616 (1960).
PENROSE, L. S.: Mongolism. Brit. med. Bull. *17:* 184–189 (1961).
RHOADES, M. M.: Preferential segregation in maize, pp. 66–80. *In* J. GOWEN, ed.: Heterosis (Iowa State College Press, Ames 1952).
SANDLER, L.; LINDSLEY, D. L.; NICOLETTI, B. and TRIPPA, G.: Mutants affecting meiosis in natural populations of *Drosophila melanogaster*. Genetics 60: 525–558 (1968).
SCHULL, W. J. and NEEL, J. V.: The effects of inbreeding on Japanese children (Harper & Row, New York 1965).

SERGOVICH, F.; VALENTINE, G. H.; CHEN, A. T. L.; KINCH, R. A. H. and SMOUT, M. S.: Chromosome aberrations in 2159 consecutive newborn babies. New Engl. J. Med. *280:* 851–855 (1969).

SIEGEL, S.: Nonparametric statistics for the behavioral sciences (McGraw-Hill, New York 1956).

STURTEVANT, A. H.: Preferential segregation in triplo-IV females of *Drosophila melanogaster*. Genetics *21:* 444–466 (1936).

THERMAN, E.; PATAU, K.; SMITH, D. W. and DEMARS, R. I.: The D trisomy syndrome and XO gonadal dysgenesis in two sisters. Amer. J. hum. Genet. *13:* 193–204 (1961).

Latent Meiotic Anomalies Related to an Ancestral Exposure to a Mutagenic Agent

K. S. LAVAPPA
GEORGE YERGANIAN

Chemical mutagens, such as nitrogen mustards, ethylenimines, and nitroso compounds induce both complete and fractional (mosaic) mutations in drosophilas (1). In several instances where expression of the mutation was delayed until the second and third generations, it was postulated that those mutagens having low energy yields may act by elevating the involved gene locus or a segment of chromosome to a labile, premutated state where it remains until additional energy is available to complete the mutational process (2). This report deals with a similar phenomenon encountered in cytologically normal F_2 generation males of the Armenian hamster, *Cricetulus migratorius* ($2n = 22$), whose grandsires were treated with a single dose of ethyl methanesulfonate (EMS). Subsequent treatment of the cytologically normal F_2 generation males with an equivalent dose of urethan (ethyl carbamate), an unrelated and virtually ineffective chromosomal mutagen for mammals (3), is interpreted as having provided the additional energy necessary for the latent expressions of two highly specific forms of meiotic chromosomal anomalies.

Details of our experiments are given in Fig. 1. Adult male Armenian hamsters were injected intraperitoneally with a single dose of EMS (100 mg/kg of body weight) in normal saline and bred on alternate days to normal females (days 1 through 11). Following unilateral orchidectomy of 15 F_1 males at 21 to 26 days of age, spermatogonial and meiotic chromosome preparations were examined for chromosomal alterations. Translocations were evident in 2 of the 15 males; the first had two tetravalents, while the other had one hexavalent with an overall incidence of 30 percent translocations among males. The 13 normal (cytologically confirmed) F_1 males and an equivalent number of female littermates were then bred to normal counterparts. Meiotic analyses of the resulting 56 F_2 male progeny derived from the 13 cytologically normal males disclosed that all F_2 males had the normal 11 bivalents (Fig. 2a). Uniform litter size and sex ratios, and failure to detect chromosomal anomalies among the 35 F_2 males derived from F_1 females, indicate that viable (multiple) translocations were restricted to two mutant F_1 males sired on the seventh and eighth days after treatment of sires with EMS (4).

The 91 unilaterally orchidectomized EMS-background "normal" F_2 progeny were included, along with 20 additional control males, in another trial to determine if urethan is, in fact, inactive throughout the meiotic cycle, as earlier reports employing mice and rats had indicated (3). Thus, the three groups of animals (EMS-background F_1 male-derived, EMS-background F_1 female-derived, and normal intact controls) were administered single intraperitoneal doses of urethan (100 mg/kg) (Merck and Co.) in normal saline. Analyses of meiotic bivalents at diakinesis were made daily for a period of 2 weeks. As in earlier reports (3), urethan was totally ineffective in the control animals

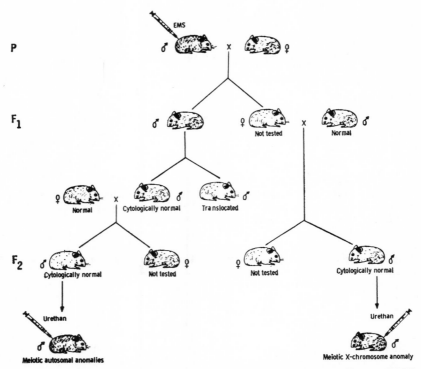

Fig. 1. Scheme of experimental procedures employed. Ethyl methanesulfonate (EMS) was administered to grandsires. F_1 progeny consisted of translocated and nontranslocated males, and females that were not analyzed for meiotic configurations. All F_2 males were cytologically normal. Certain males derived from cytologically normal F_1 animals (left side) exhibited anomalous bivalents following urethan treatment (see text). F_2 males derived from F_1 females (right side) showed mainly X-chromosome deletion following urethan administration (see text).

throughout the 14-day period of observation.

However, in sharp contrast to the negative observations on control animals, meiotic cells of particular EMS-background animals treated with urethan exhibited two distinct forms of cytological anomalies, namely, (i) association of bivalents, end-to-end, leading to a reduction in number of paired elements in diakinesis (Figs. 2, b and c), and (ii) site-specific deletions of the X chromosome at the junction of positive and negative heteropycnotic segments of the long arm (Fig. 2d) which in the somatic X chromosome is the junction of late- and early-replicating segments. Furthermore, the former anomaly was restricted to only those F_1 male-derived progeny examined on the eighth day after urethan treatment, while the distinctive X-chromosome breakage was observed in half of the F_1 female-derived progeny examined on the sixth day after urethan treatment. Observations conducted earlier and later than optimal periods for each population of animals were totally negative. Thus, events leading to either bivalent associations or specific X-chromosome lesions had taken place earlier in the reproductive cell cycle, and were detected in spermatocytes which had progressed to diakinesis by the eighth and sixth days, respectively.

Ninety percent of the cells (206/232) in four males examined on the eighth day had reduced numbers of bivalents

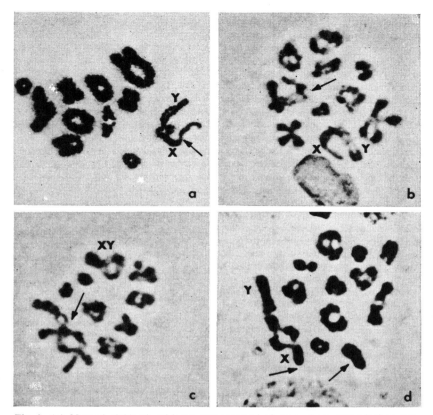

Fig. 2. (a) Normal diakinesis with ten autosomal bivalents and one XY bivalent. The latter exhibits the normal quadriradial configuration indicative of crossing-over medially on the short arms (5). Junction of positive and negative heteropycnotic segments of the long arm of the X chromosome is indicated by the arrow (air-dried preparation, lactic-acetic-orcein; × 1400). (b and c) Terminal associations of autosomal bivalents (arrows) noted 8 days after urethan treatment of F_2 males derived from EMS-background F_1 males (acetocarmine squash preparations; × 1400). (d) X-chromosome breaks noted on sixth day after urethan treatment of F_2 males derived from EMS-background F_1 females. Uniformity in size of deletion (right arrow) and length of proximal segment (left arrow) indicates site of breakage to be the junction of early- and late-replicating (positive and negative heteropycnotic) segments of the long arm (× 1900).

(Table 1). The association of two or more bivalents at terminal sites (Fig. 2, b and c) resulted in a reduction in the number of bivalent pairs from 11 to 7, 9, and 10. Occasionally, 12 bivalents were noted. Similar preparations viewed from days 1 through 7 and days 9 through 14 failed to disclose bivalent association (Fig. 3), thereby indicating the narrow time span and size of the responding sensitive cell population.

The highly specific medial breakage of the long arm of the X chromosome is, perhaps, the most impressive evidence yet encountered to suggest the possibility for later generations of mammals to express latent, premutated lesions stemming from an ancestral exposure to a weak mutagen. Two of the four males viewed on the sixth day after urethan treatment had the specific break on the X chromosome in approximately

Table 1. Meiotic anomalies noted on the eighth day of urethan treatment of F_2 males derived from F_1 male progeny. The anomalies are shown as numbers of cells of the 232 cells examined, having fewer (7, 8, 9, or 10) or more (12) bivalents than the normal number (11).

Animal	No. of bivalents						Total No. of cells in diakinesis examined
	7	8	9	10	11	12	
No. 1	1	0	3	49	9	2	64
No. 2	0	0	2	47	2	1	52
No. 3	0	0	0	46	6	2	54
No. 4	1	0	5	52	4	0	62
Totals	2	0	10	194	21	5	232

half (48/80) of the cells observed in diakinesis (Fig. 2d). Autosomal breaks were, indeed, rare, and spontaneous breakage was virtually nil in control animals. We interpret the susceptible X chromosome to have stemmed from the grandsires treated with EMS. This reasoning is based upon the following observations: (i) X-chromosome lesions were totally absent among the F_1 male-derived progeny which exhibited only associated bivalents, their X chromosome being contributed by a normal female parent; (ii) the control animals were also totally negative; therefore, urethan may be excluded as the principal mutagen; (iii) with 50 percent of the F_1 female-derived male offspring (2/4) displaying the specific X-chromosome lesion, the 1 : 1 (normal : aberrant) ratio fulfills the expected sex chromosome segregation pattern descending from EMS-treated grandsires.

Since the association and reduction in bivalents and X-chromosome breaks were present in spermatocytes at diakinesis only on specific days following urethan treatment, the mutational events had to take place earlier at particular sensitive stages of the reproductive cell cycle. Judging from the time scale for completion of the initial wave of meiosis in the Armenian hamster (5), the sensitive period for the induction of bivalent associations is narrowed to late type B spermatogonial transformation into early spermatocytes. The type B spermatogonia containing crusty chromatin in their nuclei are first noted

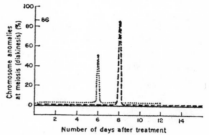

Fig. 3. Frequencies of X-chromosome breakage and association of bivalents in EMS-background F_2 males treated with urethan. Note the sharp increase in X-chromosome breakage (dotted line) and bivalent association (broken line) on the sixth and eighth days, respectively, and the lack of same in control animals (solid line) treated with urethan only. These observations are based on 2326 cells in diakinesis from 105 testes.

around the 12th day of age, reaching a plateau by the 15th day. Thereafter a sharp decrease is noted with the appearance of early spermatocytes around the 16th day. This reduction in the number of type B spermatogonia along with the appearance of early spermatocytes is referred to as the type B spermatogonial transformation. In the case of the X-chromosome breakage, the sensitive period for this induction is considered to be early spermatocytes, that is, pre-pachytene, when sex bivalents exhibit striking and distinctive patterns of paired (XY) allocycly, particularly along the long arms which are never involved in crossover configurations as displayed by the short arms of the X and Y chromosomes (5).

For the moment, a plausible mutational pathway leading to the observed site-specific deletions of the X chromosome is incomprehensible. However, structural features of the breakage site are being clarified. For example, in mitotic cells, the breakage site is the junction between the proximal, late-replicating and distal, early-replicating halves of the long arm of the active X chromosome (6). In spermatocytes, this same site is the junction between positively and negatively heteropycnotic segments (5). The latter physiological state is peculiar to the X chromosome, autosomal bivalents being relatively isopycnotic throughout meiotic prophase, and the Y chromosome gradually becoming isopycnotic following a pattern which is totally dissimilar to that exhibited by autosomes and homologous segments of the X chromosomes (Fig. 2, a and d). The manner in which the X chromosome may retain a premutable status for at least two generations is, for the moment, unclear.

Neither EMS nor urethan, administered separately or in combination to control animals, resulted in visible mutations during the immediate period (5 to 8 days) of daily examinations of spermatogonial and meiotic cells. Yet, these same agents, administered several generations apart, fostered latent expressions to become evident. Since urethan is postulated as having provided the additional energy needed to complete the mutational events initiated by EMS in late spermatids-spermatozoa of an earlier generation, it seems reasonable to describe the separate contributions of each agent leading to completion of the mutational process by the term "latent synergism."

As in the earlier findings with drosophilas (1, 2), rodents treated with weak mutagens may also express fractional mutations. However, in the case of mammals, delayed expression of fractional mutations in later generations may continue to require exposure to a relatively ineffective and unrelated agent which serves as a source of the energy needed to complete the mutaitonal process. In these trials, latent expressions were witnessed only in particular waves of transitory male germ cells. Thus, direct mutations (dominant lethals and chromosome translocations) induced by EMS tend to occur in late spermatids and epididymal spermatozoa (7, 8), whereas latent, incomplete EMS-related mutational states are witnessed only in specific clusters of midspermatocytes of later generations when exposed to urethan. The fact that spermatogenesis in mature males is associated with diurnal waves of meiotic progression means that daily exposures to weak mutagens can be expected to add measurably to the mutational load residing in epididymal sperm.

In the event that fractional mutations appear in future generations of Armenian hamsters when other combinations of agents are substituted, the merits of employing alkylating agents, in particular, for clinical therapy will have to be weighed against plausible overloading of future generations with incomplete forms of mutations. Conceivably, the ever-increasing spectrum and sophistication in the synthesis of new agents can readily serve to provide the required impetus for completion of the mutational process in a later generation. Revelation of highly specific forms of latent mutations in a laboratory rodent, by procedures described herein, is expected to add measurably toward understanding the depth and scope of the many problems related to mammalian mutagenesis.

References and Notes

1. M. L. Alexander, *Genetics* **56**, 273 (1967; C. Auerbach, *Science* **158**, 1141 (1967); *Symp. Quant. Biol.* **16**, 199 (1951); *Proc. Roy. Soc. Edinburgh Sect. B* **62**, 211 (1945); ——— and J. M. Robson, *Nature* **157**, 302 (1946).
2. C. Auerbach, J. M. Robson, J. G. Carr, *Science* **105**, 243 (1947).
3. A. J. Bateman, *Mutat. Res.* **4**, 710 (1967); C. C. Haung, *Chromosoma* **23**, 162 (1969).
4. K. S. Lavappa and G. Yerganian, *Genetics* **61**, s35 (1969). Multiple reciprocal translocations were detected in the two affected F_1 males sired on days 7 and 8 following treatment of sires with EMS. Comparison of the frequencies of reciprocal translocations induced by EMS in the Armenian hamster with 100 mg/kg body weight and that reported by Cattanach *et al.* (7) for the house mouse with a dose of 240 mg/kg body weight suggests the former species to be two and a half times more sensitive to EMS. In addition, the low number of recognizable chromosomes and meiotic bivalents of the Armenian hamster facilitates rapid identification of the translocated chromosome types.
5. K. S. Lavappa and G. Yerganian, *Exp. Cell Res.* **61**, 159 (1970).
6. G. Yerganian and S. Papoyan, *Hereditas* **52**, 307 (1965); C. Sonnenschein and G. Yerganian, *Exp. Cell Res.* **57**, 13 (1969).
7. B. M. Cattanach, C. E. Pollard, J. H. Isaacson, *Mutat. Res.* **6**, 297 (1968).
8. K. S. Lavappa, *Environ. Mutagen Soc.* No. 1 (1st Ann. Mtg.) 29 (1970).
9. We thank Henry Gagnon, James Allen, Lynn Rupert, and Inga Shields for their technical assistance. This investigation was supported in part by research grants from the National Cancer Institute (C-6516 and CA-08378), NSF (GB-7458), and the Damon Runyon Memorial Fund (293).

The Radiation and Drug Sensitivity of Meiotic Processes

PROCEDURES FOR INDUCTION OF SPAWNING AND MEIOTIC MATURATION OF STARFISH OOCYTES BY TREATMENT WITH 1-METHYL-ADENINE

MARGARET STEVENS

A polypeptide released from the radial nerve (radial nerve factor or gonad stimulating substance) induces shedding of gametes in starfish [1, 2]. In the female, this factor stimulates the ovary wall to produce a meiosis-inducing substance. This substance, which has been identified as 1-methyladenine [8], mediates both spawning and maturation of oocytes [5, 6, 7, 9]. I here describe methods for the use of 1-methyladenine (1-MeAde) to induce spawning, meiotic maturation, and subsequent development in five species of California starfish. My results confirm observations by Kanatani on the nature of the mechanism of spawning.

To examine spawning responses, ovarian fragments 2–3 mm long were washed in sea water to remove loose oocytes and then placed in small volumes (0.1–1.0 ml) of 1.3×10^{-4} M 1-MeAde (Sigma Chemicals) dissolved in sea water. To observe oocyte maturation, ovaries were torn with forceps, and the oocytes were released by agitating the tissue with a pipette. In all the species included in this study, oocytes isolated in such a manner do not mature in sea water unless 1-MeAde is present.

Six species of asteroids, including *Patiria miniata*, *Pisaster brevispinus*, *Pisaster giganteus*, *Pisaster ochraceus*, *Pycnopodia helianthoides*, and *Mediaster aequalis*, were examined for spawning of ovarian fragments, in addition to the echinoid *Dendraster excentricus* and the holothuroid *Stichopus californicus*. The non-asteroids, although ripe, did not spawn. Spawning, however, occurred in all the starfish. More extensive experiments with *P. miniata* showed that concentrations as low as 1×10^{-6} M are fully effective in inducing spawning. Spawning begins in ovarian fragments after 20–30 min incubation in 1-MeAde and continues for 5–10 min.

Using several species of starfish, Kanatani [5, 6, 7], has shown that spawning involves a reduction of intra-ovarian adhesivity due to the loss of the follicles surrounding the oocytes. These follicles are composed of numerous cells forming a hollow sphere around each oocyte. In the intact ovary each follicle adheres closely to the follicles of adjacent oocytes. 1-MeAde appears to result in dissolution of the cementing substance between the follicle cells and thus in loosening of the oocytes. Kanatani postulates that this loosening of contacts, combined with the tension already present in the ovary wall, is sufficient to force the oocytes out of the ovary.

The oocytes of the starfish used in the present study are also surrounded by follicles which dissolved as a result of treatment with 1-MeAde. In isolated oocytes of *P. miniata* dissolution occurs 20–30 min after addition of 1-MeAde, which is similar to the time required for 1-MeAde-induced spawning to occur.

Isolated oocytes incubated in calcium-free sea water for 1 h also lose their follicles. However, ovarian fragments do not spawn in calcium-free sea water, presumably because calcium is necessary to maintain the contracted state of the ovary wall [5]. If such fragments are first treated with calcium-free sea water long enough (45 min) to disintegrate the follicles and are then returned to normal sea water, spawning occurs in the absence of 1-MeAde. These findings confirm the results of similar experiments performed by Kanatani and support the mechanism of spawning that he has proposed [5]. They also indicate that this phenomenon may be widespread among starfish.

Table 1. *Germinal vesicle breakdown at various concentrations of 1-methyladenine*

Conc. 1-MeAde (M/l)	% germinal vesicle breakdown			
	Patiria miniata	Pisaster brevispinus	Pisaster ochraceus	Pycnopodia helianthoides
Sea water control	17	10	17	5
1.3×10^{-8}	28	13	24	7
1.3×10^{-7}	92	85	90	99
1.3×10^{-6}	100	93	88	100
1.3×10^{-5}	100	97	88	100
1.3×10^{-4}	100	95	91	100

Oocytes were incubated 45 min in 1-MeAde in sea water. Percentages of germinal vesicle breakdown were determined microscopically.

Table 3. *Relationship of germinal vesicle breakdown (g.v.) to fertilizability in* Patiria miniata *oocytes*

Minutes of incubation	% g.v. breakdown 30 sec aliquot	% Fertilization
0	10	9
15	16	10
20	17	29
25	40	38
30	98	98
35	100	99
40	100	100

Oocytes were incubated in 1.3×10^{-5} M 1-MeAde. The incubation times were arranged so that all samples were fertilized simultaneously. Thirty seconds before addition of sperm and again 5 min after sperm addition, aliquots were removed from each sample and fixed in 5 % formaldehyde in sea water. The 30-sec aliquot was examined microscopically to determine the percentage of maturing oocytes. Shrinkage and irregularity of the nucleus were used as criteria for germinal vesicle breakdown. The 5-min aliquot was examined to determine for each sample the percentage of fertilized oocytes, as indicated by elevation of the fertilization membrane.

1-MeAde also induced meiosis in the oocytes of five species of starfish examined (*P. miniata, P. brevispinus, P. giganteus, P. ochraceus,* and *P. helianthoides*) at concentrations as low as 1.3×10^{-7} M (table 1). At 15°C meiotic events occur in reasonable synchrony, as is reported in table 2. Twenty-five to 28 min after addition of 1-MeAde, germinal vesicle breakdown first becomes apparent as a waviness of the nuclear outline. Several minutes later the nucleolus disappears, and the nucleus begins to shrink. At 34–35 min after addition of 1-MeAde the nuclear membrane dissolves, and by 45 min no nucleus is apparent. These events were examined in detail in oocytes of

Table 2. *Timing of meiotic events in oocytes of* Patiria miniata *at 15°C*

Time in 1-MeAde (min)	State of maturation
0	1-MeAde added
25–28	Nuclear outline wavy
30–32	Nucleolus disappears; nucleus shrinks
34–35	Nuclear membrane gone
40–47	No nuclear spot apparent
95–110	First polar body
150–165	Second polar body

P. miniata and *P. helianthoides*. Preliminary observations indicate that the other species follow much the same pattern.

Delage [3] has shown that germinal vesicle breakdown must occur before starfish oocytes can be fertilized. My observations also show a chronological correlation between germinal vesicle breakdown and fertilizability in 1-MeAde-treated oocytes. In the experiment summarized in table 3, samples were incubated for varying lengths of time before addition of sperm. An aliquot from each sample was fixed 30 sec before and 5 min after sperm addition. The percentage of maturing oocytes was determined in the 30-second aliquot, and the percentage of fertilized oocytes was counted in the second aliquot. As can be seen from table 3, the greatest percentage of oocytes became fertilizable between 25 and 30 min, when the nucleus is exhibiting the first signs of breakdown. Microscopic observations also indicate that although germinal vesicle breakdown is required, the complete disappearance of the nuclear membrane is not necessary for oocytes to be fertilizable.

The results of these experiments show that 1-MeAde mediates both spawning and maturation of the oocyte. Spawning involves dissolution of the follicle surrounding each oocyte. Follicle dissolution occurs in both ovarian fragments and isolated oocytes after approximately the same incubation time. 1-Methyladenine-initiated meiosis occurs in ovaries, ovarian fragments, and isolated oocytes.

From these results the following method has been developed for obtaining starfish oocytes which yield consistently high percentages of fertilization and of normal larvae. Ovary fragments or isolated oocytes are incubated in 1×10^{-6} M 1-MeAde in sea water. Testes are removed from ripe male starfish and stored in a cool place until needed. Just before fertilization several milliliters of sea water are added to the dry testes, and the sperms are released by agitation. One drop of the resulting sperm suspension is sufficient to fertilize several milliliters of oocytes. At 15°C the oocytes are then washed to remove excess sperm and allowed to develop in a cool place. Concentrations of 1-MeAde as high as 1×10^{-3} M can be used without ill effects. A certain number of immature oocytes are unavoidable, but the number can be greatly reduced by treating ovarian fragments and collecting only those oocytes that are shed. 1-MeAde does not need to be present at the time of fertilization, as it neither activates nor inhibits the sperm.

1-MeAde can also be used for biochemical studies of meiosis, as it provides a simple way to synchronize the meiotic events of large numbers of oocytes. Studies concerning the biochemical events of meiosis are currently underway in this laboratory.

I thank Dr David Epel for his advice concerning the work reported in this paper and for his criticism of the manuscript. I also thank Dr Haruo Kanatani for a stimulating discussion of this work.

Supported by NIH Predoctoral Fellowship (5 FO1 GM41249-02) from the General Medical Sciences Institute and by NSF grant GB-8002.

REFERENCES

1. Chaet, A B, Biol bull 126 (1964) 8.
2. — Ibid 130 (1966) 43.
3. Delage, Y, Arch zool exptl gen (ser 3) 8 (1901) 284.
4. Kanatani, H, Science 146 (1964) 1177.
5. — Gen comp endocrinology, suppl. 2 (1969) 582.
6. — Exptl cell res 57 (1969) 333.
7. Kanatani, H & Ohguri, M, Biol bull 131 (1966) 104.
8. Kanatani, H, Shirai, H, Nakanishi, K & Kurokawa, T, Nature 221 (1969) 273.
9. Schuetz, A W & Biggers, J D, Exptl cell res 46 (1967) 624.

CHROMOSOMAL CHANGES IN MAIZE INDUCED BY HYDROGEN FLUORIDE GAS[1]

ALY H. MOHAMED

Introduction

Organic and inorganic fluorocompounds are potential air and water pollutants in both rural and urban environments. These compounds are emitted into the air and captured in the soil. It is thus doubtful that any unrefined natural substance is completely devoid of fluoride. Fluoride compounds are released from many industrial plants, including those producing phosphate-fertilizers, ceramics, and certain metals such as aluminum.

Gaseous hydrogen fluoride is one of the most phytotoxic of the halogen compounds occurring as air pollutants (McNulty and Newman, 1957). It is readily absorbed by leaves, but translocation to other parts of the plant is limited (Daines, Leon and Brennan, 1952). According to Adams (1956) and Ledbetter, Mavrodineanu and Weiss (1960), however, fluorides appear to have cumulative phytotoxic effect which may move through the plant vascular system to the leaf tips and margins and produce foliar necrosis. There now exists a fairly extensive bibliography on sources of fluoride contamination, injury pattern in plants, and species sensitivity to different fluoride compounds. Hill, Transtrum, Pack and Winters (1958) stated that toxic gases in the atmosphere, including fluoride, may cause severe damage to plants even when the concentration is below the level required to produce visible morphological symptoms. This is the theory of hidden or invisible injury by fluoride damage. Muller (1958) pointed out that a number of toxic substances which may occur as pollutants have been found to produce their chief damage by injuring the hereditary material of the cells they enter. He suggested that some substances, including fluorides, may produce this type of effect indirectly by giving rise to chemical reactions that result in the formation of mutagens. It has been now established that fluoride interference with the genetic system is either at the microscopic or submicroscopic level of injury. Mohamed, Applegate and Smith (1966a) reported that acqueous sodium fluoride (NaF) in a low concentration was capable of inducing chromosome breakage as well as causing chromosome stickiness and injuring the spindle to produce c-mitosis effects in mitotic cells of onion root tips. Later, Mohamed, Smith and Apple-

[1] This investigation was supported by U.S. National Air Pollution Control Administration, Public Health Service Grant AP00586.

gate (1966b) indicated that gaseous hydrogen fluoride (HF) was able to induce paracentric inversions as well as other abnormalities in meiotic chromosomes of tomato plants. Recently, Mohamed (1968) showed that the seeds obtained from tomato plants fumigated with HF, when planted, produce a number of abnormal phenotypes, the same as, or similar to, known mutants.

In view of the cytological results obtained in tomatoes, the object of the present study has been to clarify various aspects of the cytological effects on maize of HF in concentrations too low to induce visible injury. The objective was approached by analyzing the different types and degrees of meiotic abnormalities encountered in various treatments of HF.

Materials and Methods

Corn kernels of the genotype $A_1A_2C^iWx$, obtained from the Maize Genetics Cooperative, were germinated in the greenhouse in polyethylene pots containing a horticultural soil mixture. When the plants were about 2 ft high, they were placed in two growth chambers. One chamber, with 4 plants, acted as a control and the other, with 12 plants, as the fumigation chamber. The plants were left in the chambers for 2 days before starting the experiment to allow them to adjust to the environment of the growth chamber. Temperature, humidity, and light quantity were closely controlled, with a photoperiod of 14 hours, a temperature of 24°C and a relative humidity of about 60%. The concentration of the fluoride gas in the fumigation chamber was kept close to $3\mu g/m^3$. This experiment was run for 10 days with the first treated plants removed after 4 days and subsequent plant removals at 2-day intervals, thus giving a total of four treatments. Control runs were always made simultaneously with treatment runs. After each treatment period the treated and the control plants were transplanted to the field. Microsporocyte samples collected from treated and control plants were killed and fixed separately in freshly mixed ethanol propionic acid (2:1), and the material stored in 70% alcohol at 0°C until examined with propiono-carmine staining. With few exceptions, visual observations and photography were restricted to temporary slides observed immediately after preparation.

Pachytene, diakinesis, and later meiotic stages from treated and control plants were systematically scanned for chromosome abnormalities. In pachytene preparations, all cells in which the chromosomes appeared to be sufficiently spread were analyzed. Each anaphase I or II cell with poleward progress of 50% or greater was examined for the presence or absence of bridges and fragments or presence of fragments alone. Virtually all such anaphase cells were easy to classify.

Results and Discussion

At the time the microsporocytes were collected, the fumigated plants from all treatments were healthy and showed no visible symptoms of fluoride injury. Cytological studies, however, indicated the occurrence of chromosomal abnormalities.

While the pachytene chromosomes of both the control and the treated plants showed asynaptic regions, there was a much higher frequency in the treated (Table I). Also, the percentage of asynapsis increased with the treatment duration. This indicates that HF has a cumulative effect as was pointed out by Adams (1956) and Mohamed et al. (1966b). These asynaptic regions

TABLE I

Pachytene pairing frequencies in percentages

Treatment in days	No. cells	Asynaptic regions			Total Abnorm.
		1	2	>2	
Control	72	2.8	1.4	—	4.2
4	122	18.8	2.4	3.3	24.6
6	94	15.9	10.6	13.8	40.4
8	106	16.0	14.2	19.8	50.0
10	64	29.7	12.5	9.4	51.6

were either intercalary (Fig. 1) or terminal (Fig. 2). These regions might be the result of heterozygous structural changes, such as deficiencies or duplications, or might result from the induction of asynaptic mutant genes. Recessive asynaptic mutants are known to have been induced in many plants, such as maize (Beadle 1930) and pepper (Morgan 1963). In many of the preparations, chromosome 9 showed terminal asynapsis of the short arm (Fig. 3), similar to that observed in chromosome 8 by Miller (1963) in his studies on the *as* gene in maize.

Pachytene studies also indicated the presence of heterozygous inversions (Fig. 3) and translocations (Fig. 4). Accordingly, the frequencies of asynapsis in the present studies were based on cells showing neither translocations nor inversions since such structural changes are known to cause mechanical failure of chromosome pairing.

Studies at diakinesis (Table II) showed a high frequency of univalents and rod bivalents with the increase in treatment duration. Table II shows that the mean bivalent frequency in the treatment, except for 10 days, did not differ from that of the control. However, the total aberrations (Table II) clearly increased over the control, with an accumulative effect. Since no lack of complete pairing was noticed in pachytene, then complete or partial terminalization of the chiasmata by diakinesis would account for these findings (Moffett, 1932; Miller, 1963). The formation of univalents or rod bivalents might indicate that HF treatment had an effect on chiasma formation or on the strength of terminalization. No equational division of univalents during division I was observed. The nondividing univalents appeared to have moved without apparent centromere activity to either of the two poles. Such movement was precocious with regard to the remaining bivalents in metaphase I.

Figs. 1-7. Microsporoctye smears fixed in 1:2 propiono-alcohol and stained with iron propiono-carmine. Magnification ca 1350 ×. Fig. 1. Pachytene chromosomes showing terminal asynaptic regions (indicated by arrows). Fig. 2. Pachytene stage showing chromosome breakage and chromosome 10 showing intercalary asynapsis (indicated by arrows). Fig. 3. Pachytene stage showing a heterozygous inversion; also showing a terminal asynaptic region for chromosome 9 (indicated by arrows). Fig. 4. Pachytene chromosomes showing the presence of heterozygous reciprocal translocation (indicated by an arrow). Fig. 5. Diakinesis showing a 'pseudo-quadrivalent' (indicated by an arow). Fig. 6. Anaphase I with bridges and fragments. Fig. 7. Chromosome 10 showing a ring for one chromosome but non-attached for the other homolog (indicated by three arrows).

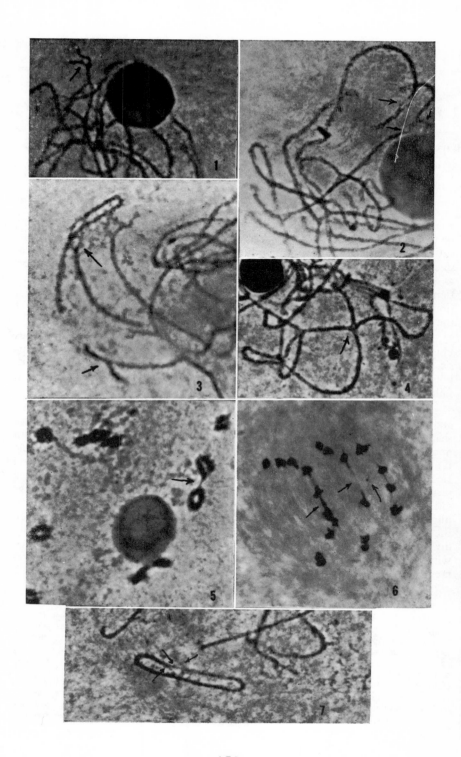

TABLE II
Frequency of abnormal diakinesis association

Treatment in days	No. cells	X̄ bivalents	Rod bivalents			Univalents	Ring of four	Total abnorm.
			1	2	4	2		
Control	144	9.91	10	1	—	—	—	11
4	184	9.83	15	5	—	—	3	23
6	193	9.68	33	5	—	8	5	51
8	133	9.65	29	5	—	7	—	41
10	118	9.23	29	21	6	5	3	64

Diakinesis also showed the presence of loosely associated bivalents, forming 'pseudo-multivalents' (Fig. 5). However, by metaphase I most such associations seemed to come to an end resulting in the formation of separate bivalents. Such association may merely be due to chromosomal stickiness rather than actual chromosomal changes. As indicated by Mohamed et al. (1966b) hydrogen fluoride may have a physiological or primary effect that causes the chromosomes to be sticky. This stickiness did not prevent anaphase movement. However, in plants No. 5a-11 (10-day treatment), a high frequency of quadrivalents was observed in metaphase I with anaphase I showing 20 chromosomes in each pole. This suggested that the spindle formation might have been affected in the pre-meiotic cycle. Such polyploid cells were observed in onion root tip chromosomes after NaF treatment (Mohamed et al. 1966b).

The formation of bridges plus fragments, or fragments by themselves, in anaphase I and II (Fig. 6) could be due to the physiological effect of HF, i.e. stickiness of the chromosomes, to the presence of heterozygous paracentric inversions, or to both as previously reported by Mohamed et al. (1966b), and Mohamed (1968, 1969). However, the presence of bridges and fragments in anaphase does not necessarily indicate the presence of heterozygous inversions.

TABLE III
Percentage of cells with chromosomal aberations in anaphases and telophases of the first and second meiotic division

Treatment in days	Anaphase I + telophase I				Anaphase II + telophase II			
	No. cells	Bridges + fragments	Frag.	Total aberr.	No. cells	Bridges + fragments	Frag.	Total aberr.
Control	135	1.5	4.4	5.9	127	2.4	2.4	4.8
4	125	4.8	15.2	20.0	107	—	4.7	4.7
6	135	14.8	8.9	23.7	162	5.6	8.6	14.2
8	72	14.0	9.7	23.7	105	20.0	10.5	30.5
10	63	23.8	11.1	34.9	151	16.6	19.2	35.8

TABLE IV
Percentages of pollen sterility

Treatment in days	Pollen counted	Sterile	Range of sterility
Control	670	8.1	6.2 - 14.8
4	853	34.1	8.1 - 59.6
6	870	38.9	7.2 - 97.8
8	940	44.3	7.0 - 98.1
10	839	48.2	15.2 - 100

According to McClintock (1941), chromatid breakage and reunion may cause the formation of bridges and fragments. Such breaks were observed in the present studies (Figs. 2,7). Miller (1963), and Baker and Morgan (1969) attributed the presence of fragments in their studies on the *as* gene in maize to misdivision of univalents. This could also be another source of fragments since univalents were observed in the present studies. Furthermore, it is known that HF affects enzymatic activity in plants (Adams 1956). This effect might delay the meiotic cycle. According to Rees (1952, 1962) and Darlington and Haque (1955), microsporocytes delayed in meiosis can exhibit asynapsis or chromosomal breakage. Such chromosomal breakage may eventually form structural changes and fragmentation.

There was also a marked increase in chromosomal aberrations with an increase in treatment duration (Table III). While the treatment for 10 days had the highest percentage of total aberration, the increase was not as sharp as that pointed out by Mohamed *et al.* (1966b) in the effects of HF on tomato chromosomes. However, the increment of the increased aberrations was 6% in each treatment duration, indicating a further accumulative effect of HF. Pollen counts (Table IV) showed an increase in pollen sterility with treatment duration, indicating the cumulative effect of HF. It was noticed that there was a great variability, as well as differences of expression, within cells of the same anther. Such variability was attributed to the stage of development at which the meiocyte cells were subjected to HF fumigation. Also, environmental effects should not be excluded.

Although no failure of cytokinesis was reported in the earlier studies of the author (1966a, 1966b, 1968) partial failure of cytokinesis was frequently observed in the present studies. Such failure of cytokinesis after karyokinesis will give rise to diploid spores. Such findings were similar to those reported by Miller (1963), and Baker and Morgan (1969).

The present studies showed clearly that hydrogen fluoride gas at a concentration below that causing any visible injury was able to induce permanent chromosomal changes and probably to induce asynaptic mutants. Therefore, this gas can be considered to have a mutagenic effect.

References

Adams, D. F. 1956. The effects of air pollution on plant life. Am. Med. Assoc. Arch. Ind. Health **14**: 229-245.
Baker, R. L., and Morgan, D. T., Jr. 1969. Control of pairing in maize and meiotic interchromosomal effects of deficiencies in chromosome 1. Genetics **61**: 91-106.

Beadle, G. W. 1930. Genetical and cytological studies of Mendelian asynapsis in *Zea mays*. Cornell Univ. Agr. Exp. Sta. Mem. **129**: 1-23.

Daines, R. H., Leone, I. A., and Brennan, E. G. 1952. The effect of fluoride on plants as determined by soil nutrition and fumigation studies, Chapter 9. *In* L. C. McCabe [ed.] Air pollution. McGraw-Hill, New York.

Darlington, C. D., and Haque, A. 1955. The timing of mitosis and meiosis in *Allium ascalonicum*: A problem of differentiation. Heredity **9**: 117-127.

Hill, A. C., Transtrum, L. G., Pack, M. R., and Winters, W. S. 1958. Air pollution with relation to agronomic crops: VI. An investigation of the hidden injury theory of fluoride damage to plants. Agron. J. **50**: 562-565.

Ledbetter, M. C., Mavrodineanu, R., and Weiss, A. J. 1960. Distribution studies of radioactive fluorine-18 and stable fluorine-19 in tomato plants. Contrib. Boyce Thompson Inst. **20**: 331-348.

McClintock, B. 1941. The stability of broken ends of chromosomes in *Zea mays*. Genetics **16**: 175-190.

Miller, O. L., Jr. 1963. Cytological studies in asynaptic maize. Genetics **48**: 1445-1466.

Moffett, A. A. 1932. Chromosome studies in Anemone. I. A new type of chiasma behavior. Cytologia **4**: 26-37.

Mohamed, A. H. 1968. Cytogenetic effects of hydrogen fluoride treatment in tomato plants. J. Air Pollution Control Assoc. **18**: 395-398.

Mohamed, A. H. 1969. Cytogenetic effects of hydrogen fluoride on plants. Fluoride **2**: 76-84.

Mohamed, A. H., Applegate, H. G., and Smith, J. D. 1966a. Cytological reactions induced by sodium fluoride in *Allium cepa* root tip chromosomes. Can. J. Genet. Cytol. **8**: 241-244.

Mohamed, A. H., Smith, J. D., and Applegate, H. G. 1966b. Cytological effects of hydrogen fluoride on tomato chromosomes. Can. J. Genet. Cytol. **8**: 575-583.

Morgan, D. T., Jr. 1963. Asynapsis in pepper following X-irradiation of pollen. Cytologia **28**: 102-107.

Muller, H. J. 1961. Do air pollutants act as mutagens? *In* Symposium on emphysema and chronic bronchitis syndrome (Aspen, Colorado, June 13-15, 1958). Am. Rev. Respirat. Diseases **83**: 571-572.

Rees, H. 1952. Asynapsis and spontaneous chromosome breakage in Scilla. Heredity **6**: 89-97.

Rees, H. 1952. Developmental variation in expressivity of genes causing chromosome breakage in rye. Heredity **17**: 427-437.

Meiotic Consequences of a Combined Treatment Fast Neutrons-FUDR in *Vicia faba*

Some chemicals, like the bifunctional alkylating agent Myleran and the pyrimidine analogue 5-fluorouracil deoxyriboside (FUDR), have been reported to increase the effects of ionizing radiations in post treatment [1-4].

The synergistic effect obtained is higher after Co^{60} γ-rays than after fast neutrons [2-4]. Although the theory has been advanced that FUDR exerts its chromosome breaking effect by inhibition of the enzyme thymidilate synthetase, the mechanism by which the effects of radiations are modified is not yet clear but should be different from the previous one. At some Co^{60} γ-rays doses, it was found that all classes of chromosome aberrations are enhanced. Since it is well known that neutron effects are far more difficult to modify, see e.g. oxygen effect, it was in the scope of the present experiment to see if treatments by FUDR after fast neutron irradiations result in observable effects at meiosis.

Material and technique. *Vicia faba* dry seeds ssp. *minor* from Gembloux (Belgium) were irradiated by fast neutrons (fission) in the following experimental conditions: ITAL reactor (Wageningen, the Netherlands). Reactor power: neutron flux density was 2.10^7 $n/cm^2/sec$; neutron fluence applied was $2.2 \cdot 10^{10}$ n/cm^2 corresponding approximately to an absorbed neutron dose of 125 rad; γ contamination was about 140 rad/h. After irradiation, half of the seeds were treated by a solution of FUDR (concentration: 0.1 mg/100 ml). A control set (not irradiated) was also treated by FUDR and compared with a untreated set. All seeds were sown on perlite medium.

After 2 weeks, seedlings were transplanted in liquid medium (Hoagland) under bubbling conditions. They were grown in the following conditions: light intensity: 15.000 lux; photoperiod: 16 h light. Under these conditions they reached the flowering period in 2 months. At that time flower samples were collected for cytological investigation (5 samples in each series). They were fixed with Carnoy (2 h) then transferred into 70° alcohol. They were stained according to Feulgen technique and mounted in Depex.

Results. The criteria on which the present analysis is based have different meanings according to the stage investigated. The stages going from diplotene to metaphase I are generally suitable to identify translocations. Anaphase I allows us to observe the consequence of translocations and also other structural changes. Mitosis II yields mostly information about the selection of the aberrations from which some inference can be made on the nature of these aberrations.

In Tables I and II, all types of aberrations have been put together.

Table I. Proportions of aberrations observed in different meiotic stages (5 different samples put together in each case)

Stages	Diplotene		Diakinesis		Metaphase I	
	No. of cells analysed	No. of abnormal cells	No. of cells analysed	No. of abnormal cells	No. of cells analysed	No. of abnormal cells
Control	150	0	200	0	700	0
FUDR	120	0	200	0	900	3
Neutrons	94	3	64	0	548	2
Neutrons + FUDR	93	3	116	2	387	13

Table II. Proportions of aberrations observed at different meiotic stages (5 different samples put together in each case)

Stages	Anaphase I		Anaphase II		Tetrads	
	No. of cells analysed	No. of abnormal cells	No. of cells analysed	No. of abnormal cells	No. of cells analysed	No. of abnormal cells
Control	400	1	1000	1	3000	3
FUDR	430	6	1200	9	5000	13
Neutrons	300	26	1260	13	2500	23
Neutrons + FUDR	344	90	862	31	1450	32

Table III. Proportions of different kinds of aberrations at anaphase

		Chromosome bridges	Chromatid bridges	Fragments	Attached fragments	Bridges + fragments	Micro-nuclei	Dissociation of trans-locations
Anaphase I	Neutrons	1	7	4	4	6	2	2
	Neutrons + FUDR	11	22	20	28	5	2	2
Anaphase II	Neutrons	–	4	6	–	2	1	–
	Neutrons + FUDR	–	12	15	–	2	2	–

The bar indicates that the category does not exist.

No significant difference appears for diplotene and diakinesis at which some aberrations can remain unnoticed. On the other hand, there is a significant increase in the amount of aberrations scored at metaphase I. The aberrations observed are mostly rings of 4. No significant difference in the relative proportion of rings of 4 and figures of 8 exists for the 2 series, neutrons and neutrons + FUDR. The amount of aberrations induced by FUDR alone is low. The difference when FUDR is added is more striking for anaphases I and II as well as for resulting micronuclei observed in the tetrads (Table II).

Some aberrations at anaphase I are clearly identified as the consequence of translocations like the above-mentioned rings of 4 and figures of 8. The occurrence of attached fragments is somewhat more difficult to interpret. They are generally considered as subchromatid aberrations. When the possibility of an artefact is ruled out, the only explanation could be based on the existence of a delayed effect which is far higher when FUDR is added.

A comparison of the amount of aberrations is given for different types in Table III. Some are clearly of other origins than the ones mentioned above and were not detected at early stages. At anaphase I, some classes of aberrations are not increased. This is specially the case for bridges with 1 fragment, the probable origin of which being some kind of inversions.

For the series neutrons + FUDR, there is a selection of some aberrations from the first to the second mitosis although the figures remain still higher than for neutrons alone (Table II).

The distribution of the aberrations seems to be the same at anaphase II (Table III) since bridges and fragments are almost in equal numbers.

Conclusions. This research confirms the possibility of modifying the effects of fast neutrons by FUDR in post treatment. Moreover it shows that this effect results in a higher amount of aberrations at meiosis.

It should be pointed out that there is no difference in induced zygotic sterility between the two series, but this last statement should be extended to higher neutron fluences.

An increased sterility might possibly be entirely on the gametic side. Experiments designed to analyse $X2$ generations for the mutation rate are being carried out[5].

[1] M. MOUTSCHEN-DAHMEN, J. MOUTSCHEN and L. EHRENBERG, Radiat. Bot. *6*, 251 (1966).
[2] M. MOUTSCHEN-DAHMEN, J. MOUTSCHEN and L. EHRENBERG, Radiat. Bot. *6*, 425 (1966).
[3] J. MOUTSCHEN and N. DEGRAEVE, Experientia *22*, 581 (1966).
[4] J. MOUTSCHEN, M. K. JANA and N. DEGRAEVE, Caryologia *19*, 4 (1966).
[5] This work was aided by Euratom-I.T.A.L. The authors wish to express their gratitude to Drs. DE ZEEUW, ECOCHARD, CONSTANT and DE NETTANCOURT for their great interest in this research. They are also indebted to Mrs. N. LABOURY and Miss M. T. MUD for their enthusiastic and skilful assistance.

The Action of Phleomycin on Meiotic Cells[1]

Yasuo Hotta and Herbert Stern

INTRODUCTION

The distinctive action of phleomycin in blocking the replication of HeLa cells without appreciable effect on DNA synthesis has been fully described by Kajiwara et al. (4). These authors concluded that the drug acts on the G_2 phase of the cell cycle and thus prevents cells from entering the prophase of mitosis. They also attempted to reconcile their observations on G_2 arrest with the fact that phleomycin can inhibit DNA synthesis in *E. coli* (7) by speculating that the primary action of the drug is on chromosome structure and that the particular consequence of such action may depend upon the amount of DNA within that structure. Additional evidence pointing to a direct effect of phleomycin on chromosome structure was provided by Kihlman et al. (5) who found that low concentrations of phleomycin, which affected neither DNA synthesis nor mitosis in root tips of *Vicia faba* produced various chromosomal aberrations.

---

[1]This investigation was supported by NSF Grant GB 5173X and supplemented by USPHS Grant HD03015 from the National Institute of Child Health and Human Development.

The purpose of this communication is to provide additional evidence on the mode of phleomycin action based upon observations of cultured meiotic cells. Although the effects of this drug on meiosis may be interesting on their own account, meiotic cells permit a more refined analysis of phleomycin effects on chromosome behavior. The interval between the premeiotic S phase and the completion of division is about 10–12 days. The G_2 period is about 1–2 days, and the prophase period (leptotene, zygotene, pachytene, and diplotene), during which the chromosomes undergo a sequence of morphologic changes, extends for about 8 days. The meiotic cycle is punctuated by three distinctive periods of DNA synthesis: the first occurring in premeiosis and corresponding to the conventional S phase of mitotic cells, the second occurring at zygotene in association with chromosome pairing, and the third occurring at pachytene in association with what is generally believed to be chromosome exchange (3). The question may thus be asked as to whether phleomycin has any selective effect on three different types of DNA synthesis. The protracted meiotic prophase, by contrast with the rapid mitotic prophase, makes possible an examination of phleomycin action on chromosome structure during an interval when the structure is undergoing significant modifications. These studies are facilitated by the fact that the meiotic cells used become synchronous in development following the G_2 interval.

The data reported here confirm the conclusions of Kajiwara *et al.* (4). However, they reveal two additional aspects of phleomycin action which were not evident in studies of mitotic cells: (*a*) The drug not only inhibits the entry of cells from the G_2 phase into mitosis and meiosis, but it also blocks the progress of cells at almost any stage following their entry into meiosis; (*b*) The drug promotes a gradual and cumulative change in chromosome structure which is manifested as an intranuclear condensation of chromosomal material.

MATERIALS AND METHODS

Meiotic cells from "Cinnabar" and "Bright Star" lilies were explanted and cultured as previously described (6). Phleomycin (Lot No. A9331-648) was obtained from Bristol Laboratories, Syracuse, New York. Following exposure to phleomycin for the period indicated, the cells were washed and resuspended in fresh culture medium. "Cinnabar" cells were generally cultured at 20°C and those of "Bright Star" were cultured at 15°C. For cytologic studies, cells were fixed in ethanol:acetic acid (3:1 v/v) and stained with propionic orcein. Some preparations were stained with the Feulgen reagent and a few were stained for histones by the fast green stain of Alfert and Geschwind (1). In reporting percentages of cells affected by a particular treatment, at least 1000 cells were counted in each preparation. The procedures used in labeling and characterizing DNA were the same as those reported earlier (2). In all experiments, the cells obtained from any single flower bud were divided into two portions one of which was retained for control cultures.

RESULTS   DNA Synthesis

The effects of phleomycin on the three intervals of DNA synthesis associated with the meiotic cycle are shown in Chart 1. Cells in the premeiotic S phase and those in zygotene behave similarly; no inhibitory effect is evident at concentrations of 20 μg/ml or less. DNA synthesis in the meiotic cells appears to be somewhat less sensitive to phleomycin than in the HeLa cells, but, as will be seen, both cell types are inhibited in division by concentrations of the drug which do not affect DNA synthesis. A differential response is, however, clearly evident if zygotene and pachytene cells are compared. DNA synthesis at the pachytene stage is more sensitive to phleomycin. The higher sensitivity does not reflect a fixed reduction in rate of DNA synthesis. This is shown by the curves for pachytene synthesis in Chart 1. The upper curve was obtained from cells which had been exposed for 2 days to phleomycin and tritiated thymidine, whereas the lower curve was obtained from the same cells following removal of the medium containing thymidine and phleomycin and replacing it with fresh medium containing phosphate-^{32}P. The progressive decline in synthetic ability of pachytene cells following exposure to phleomycin is evident from a comparison of the two curves. Since the amount of pachytene synthesis is too small to make meaningful quantitative comparisons between controls and treated cells over shorter intervals of time (2), one can only speculate that the initial effect of phleomycin may not be on DNA synthesis and that the observed decline is due to other changes in the chromosome.

Since the three periods of DNA synthesis during the meiotic cycle have qualitatively different patterns, the DNA synthesized during each of the periods in the presence of phleomycin was analyzed by centrifugation in a cesium chloride gradient. The results showed no appreciable differences between treated and control cells. Thus, to the extent that each of the synthetic periods represents a distinctive physiologic function, low concentrations of phleomycin would appear to have no effect on the composition of the DNA synthesized even when, as in pachytene, a lowering of the rate occurs. The persistence of stage-characteristic differences is illustrated in Chart 2. In this experiment zygotene cells were exposed simultaneously to phleomycin and thymidine-^{3}H for two days and then transferred to fresh medium containing phosphate-^{32}P for an additional two days. The distribution of label along the cesium chloride gradient was the same as that observed in the controls (2). The tritium label is found in the heavy portion of the gradient which is characteristic for zygotene, and the ^{32}P label is in the light portion as is characteristic for pachytene. This experiment, however, provides one additional piece of information concerning phleomycin action. Under the conditions used, untreated zygotene cells complete their synthesis within

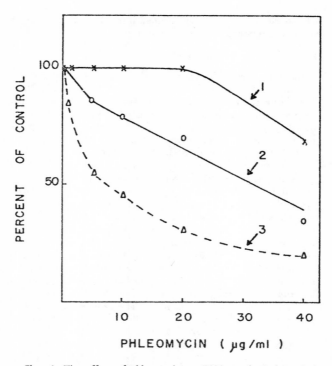

Chart 1. The effect of phleomycin on DNA synthesis in meiotic cells. Curve 1 represents the premeiotic S phase and the zygotene stage. Both stages showed the same response. Curves 2 and 3 represent the pachytene stage. DNA synthesis was measured by tritiated thymidine or phosphate-^{32}P incorporation over a period of two days. In the case of the pachytene stage, a double labeling experiment was done. Curve 2 represents the synthesis in pachytene cells exposed to tritiated thymidine and phleomycin. Curve 3 represents the incorporation of phosphate-^{32}P into DNA of these same cells after their removal from the thymidine and phleomycin medium and continued culture in drug-free medium for an additional 2 days. Curve 3 thus illustrates a decline in DNA synthesis in pachytene cells following two days of exposure to phleomycin. Isotope concentrations: phosphate-^{32}P (carrier free), 5 µc/ml; thymidine-^{3}H (4 c/mmole), 10 µc/ml.

two days and then enter into the pattern of synthesis typical for pachytene. The exposed cells behave in the same way even though they show a reduced rate of synthesis for the pachytene stage. The fact that zygotene cells exposed to phleomycin can subsequently initiate a pachytene type of synthesis is of considerable interest, since such cells do not progress into the pachytene stage as determined by cytologic analysis.

The results thus point in one direction. Low concentrations of phleomycin do not alter normal patterns of DNA synthesis in any appreciable way. Moreover, analysis of absorption spectra and of thermal denaturation patterns reveal no differences between controls and treated cells. Phleomycin also shows no appreciable influence on rates of RNA or protein synthesis during the first three days of cell exposure (Table 1). Superficially, at least, macromolecular synthesis in meiotic cells would appear to be unaffected by the presence of phleomycin.

Progress of Cells through the Division Cycle

Premeiotic cells exposed to phleomycin fail to enter mitosis or meiosis. This experiment is made possible by the fact that cells explanted in the S phase or early G_2 phase revert to a mitotic type division, whereas those explanted in late G_2 invariably enter meiosis (3). Since synchrony within a population of cells from a single anther is still incomplete during this interval, explants usually contain both types of cells. The typical results obtained when premeiotic cells in the S or G_2 phases are exposed to phleomycin are summarized in Table 2. The treated cultures show no mitotic divisions nor any leptotene forms. The action of phleomycin may be dissociated from any effects the drug might have on DNA synthesis not only because such low concentrations of phleomycin show no effect on DNA synthesis but also because the cells which would otherwise enter leptotene were exposed to phleomycin following completion of the S phase. The inhibitory action of phleomycin is thus an immediate one at the G_2 stage and is not dependent upon any prior action of the drug during the interval of chromosome replication. This action occurs with concentrations, at least, as low as 1.0 μg/ml. Exposure periods as short as three hours are sufficient to assure the inhibitory action for at least 10–12 days during which period the treated cells remain viable. Thus, under the conditions tested, the inhibitory effect is immediate and irreversible.

This action of the drug is not, however, confined to the G_2 interval. Cells exposed briefly or continuously to phleomycin at any meiotic stage, at least up to metaphase, fail to advance to the succeeding stage. Instead, such cells gradually develop characteristic cytologic abnormalities, the nature of which is described below. The time at which these abnormalities become apparent in a particular cell is variable, but the cytologic arrest is uniform for all cells. Phleomycin arrests cells at any stage during the extended prophase of meiosis; however, this

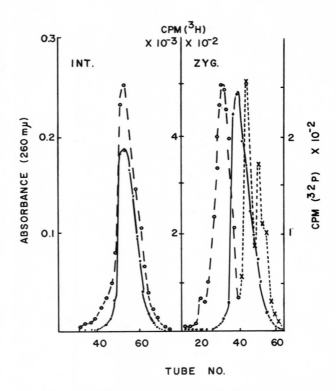

Chart 2. The density distribution of DNA synthesized in the presence of phleomycin during interphase and zygotene. The labeled DNA prepared from cells at each of the stages was centrifuged in a cesium chloride solution to an equilibrium position. The *solid lines* represent total DNA of the preparation as measured by absorbance at 260 mμ. This DNA serves as a marker since the peak concentration consistently has a density of 1.702 gm/ml. The *dashed lines* represent radioactivity. At the premeiotic S phase, the radioactivity tracks the optical density and is identical in pattern with untreated cells. Zygotene cells were double labeled. The *long-dashed lines* represent the distribution of radioactivity during the first two days of exposure to tritiated thymidine. The *short-dashed line* represents the distribution of radioactivity in the cells transferred from the thymidine medium and incubated for an additional two days in the presence of phosphate-^{32}P. The *heavy* and *light peaks* are characteristic for normal zygotene and pachytene synthesis respectively (2). In all experiments, the cells were exposed to a phleomycin concentration of 2 μg/ml. INT, interphase; ZYG, zygotene.

conclusion may not apply to metaphase and subsequent stages. At these later and much briefer stages, cytologic abnormalities also appear, but these might have been generated in the course of actual progression through the cycle. Thus, with the exception of this single qualification, the general conclusion may be drawn that phleomycin acts immediately and irreversibly to inhibit the progression of structural changes which normally occur in meiotic chromosomes and that such inhibition can be effected at concentrations of the drug which produce no identifiable effects of macromolecular syntheses.

Cytologic Effects

In cells which have been exposed to phleomycin, even for as little as three hours, the chromosomes gradually undergo a progressive condensation. At first one or a few sites of condensation become evident. These gradually increase in number and/or size. Eventually, a single large region of condensation is formed, presumably by a fusion of the smaller ones. These regions stain intensely with orcein, Feulgen, or fast green indicating that these regions include both DNA and protein. As the condensed regions increase in size, the remainder of the nucleus stains less intensely. The nucleolus appears to be unaffected. The cytologic appearance of the cells at various stages of chromosome condensation is shown in Figs. 1 and 2.

The condensation itself, as manifested under the light microscope, cannot be the cause of arrest in mitotic or meiotic development. This conclusion is based upon two pieces of

Table 1

Stage	Phleomycin (µg/ml)	DNA-^{32}P (cpm/µg)		RNA-^{32}P (cpm/µg)		Protein-leucine-^{3}H (cpm/20 µg protein)	
		24 hr	72 hr	24 hr	72 hr	24 hr	72 hr
Late S phase	0	151	565	1982	7400	421	1097
	2	149	568	1870	7487	410	1112
Zygotene	0	7.5	28.1	792	1682	105	315
	2	7.5	28.8	771	1703	98	301
Pachytene	0	4.8	14.1	409	1247	47	96
	2	3.1	10.2	386	1118	42	98

Effect of phleomycin on rates of nucleic acid and protein synthesis in meiotic cells. Meiotic cells of Cinnabar were explanted into culture media in the presence or absence of phleomycin at the stages indicated. The isotopic compounds were used at the following concentrations: phosphate-^{32}P (carrier-free), 5 µc/ml; leucine-^{3}H (1.2 c/mmole), 5 µc/ml. Cells were withdrawn for analysis after 24 and 72 hours of culture. The numbers represent specific activities.

evidence. First, premeiotic interphase nuclei, including those in late G_2, do not show any condensed regions even after 10 days in culture. Yet, as shown in Table 2, these cells do not enter meiosis. Second, the frequency and rate of appearance of condensed regions vary with the stage at which the cells were exposed to the drug, but meiotic progress is equally inhibited at all stages. Moreover, within any single culture the arrest is uniform regardless of whether the cells do or do not show condensed regions. A summary of results for six different cultures is given in Table 3. The cells are classified cytologically as Types I, II, or III. Type I includes nuclei which are normal in appearance and also those which appear slightly abnormal but which show no chromosomal condensations. Types II and III represent early and later stages of chromosome condensation as illustrated in Fig. 1.

Cells at any of the stages from leptotene to midpachytene show a progressive shift towards a preponderance of highly condensed nuclei (Type III) with passage of time. There are two sharp transitions in cell behavior which are not evident from a reading of Table 3. The first occurs at the initiation of leptotene and the second at late pachytene. Leptotene cells manifest nuclear condensations, whereas those immediately prior to leptotene do not. Cells from late pachytene on generally do not retain a coherent nuclear structure. The chromosomes appear to condense individually or in groups, and such condensation is accompanied by irregular cytokinesis and extensive wall synthesis (Fig. 2). Three general conclusions may be drawn from these observations of cytologic appearance: (a) The principal, if not the sole, observable effect of phleomycin action is on chromosome structure. (b) The appearance of gross structural lesions depends upon the particular organization of the chromosomes at the time that the cell is exposed to the drug. (c) The primary effect of phleomycin is to inhibit organizational changes within the chromosome such as those required for the normal progression of events in the meiotic cycle.

Table 2

Culture No.	Phleomycin	Stage distribution (%)					
		6 days			10 days		
		Inter.[a]	Mit.	Lept.	Inter.[a]	Mit.	Lept.
1	−	40.5	5.2	54.3	60.9	2.1	37
	+	100			97.3		2.7
2	−	98.8	1.2		99.1	0.9	
	+	100			100		
3	−	77.8	1	21.2	88	2.5	9.5
	+	100			99.7		0.3

Effect of phleomycin on premeiotic cells. Premeiotic cells of Bright Star were explanted into culture media and exposed for 1 day to phleomycin (2 μg/ml). At the time of explantation, all cells were in either the S or G_2 stages of premeiotic interphase. Cultures were incubated at 15°C and sampled after 6 and 10 days. All cells in Culture 2 reverted to mitotic division, whereas an appreciable fraction of cells in Cultures 1 and 3 entered meiosis. Cells in Culture 2 were either in S-phase or early G_2; cells in Cultures 1 and 3 were in G_2.

[a] Inter., premeiotic interphase; Mit., mitosis (prophase, metaphase, or anaphase); Lept., leptotene. Cells counted as leptotene also included the zygotene stage in 10-day-old cultures.

Table 3

	Expt. No.	Phleomycin (μg/ml)	Exposure time (hr)	Initial stage	Culture period (days)	Cytological types (Percent)			Control stage
						I	II	III	
Cinnabar	1	2.0	48	Leptotene-Zygotene	2	10	68	22	Zygotene
					8	3	17	80	Division II:Tetrads
	2	0.5	48	Leptotene-Zygotene	2	58	31	11	Leptotene-Zygotene
					8		75	25	Pachytene
	3	2.0	48	Zygotene	2	4	82	14	Early pachytene
					8	2	23	75	Tetrads
	4	2.0	48	Early Pachytene	2	3	88	9	Pachytene
					8	4	17	79	Tetrads
	5	2.0	48	Pachytene	2	36	50	14	Pachytene
					8	20	7	73	Tetrads
	6	2.0	48	Late pachytene	2	70	27	3	Pachytene
					8	17	39	44	Tetrads
Bright Star	7	1.0	3	Leptotene-Zygotene	6	15	5	80	Zygotene
					13		6	94	Pachytene

Cytologic effects of phleomycin on meiotic cells. In each experiment meiotic cells explanted from a flower bud were divided into 2 groups, one of which was cultured in the absence of phleomycin. Cinnabar cells were cultured at 20°C, and Bright Star cells were cultured at 15°C. The last column indicates the cytologic stage of the control cultures at the end of the culture period. In all cases, cells exposed to phleomycin did not advance beyond the stage at which they were first exposed to the drug. The nature of the different cytologic types is described in the text and illustrated in Figs. 1 and 2.

DISCUSSION

The most important conclusion which can be drawn from these studies is that the primary effect of phleomycin is on some component of the cell which, directly or indirectly, regulates changes in chromosomal organization. The particular type of change, whether from interphase to mitotic prophase, from leptotene to zygotene, or from zygotene to pachytene, appears to be incidental. In all cases phleomycin blocks the progression of chromosomes from one organizational form to another. The direction of change, whether from the extended to the condensed form or *vice versa*, is also incidental to phleomycin action as far as can be judged from the behavior of meiotic cells. The return to an interphase condition from the diplotene stage is as effectively blocked as condensation from the premeiotic interphase. An important question which cannot be answered from these studies is whether phleomycin action is restricted to those changes in chromosome contraction and extension which are characteristic of nuclear division cycles, or whether it also extends to localized changes such as heterochromatinization or even puffing.

Since the molecular basis of chromosome contraction and extension is unknown, it is idle to speculate about the mechanism of phleomycin action. The nature of phleomycin action is, nevertheless, distinctive inasmuch as it can arrest cell development prior to any detectable change in macromolecular syntheses. Changes in biosynthetic activity must eventually occur following phleomycin exposure, but we have not analyzed for such changes except in pachytene cells which showed a progressive decline in the rate of DNA synthesis. The relevance of delayed metabolic changes to the primary action of the drug is doubtful, since the fate of a cell appears to be sealed after no more than three hours of exposure. The effect of phleomycin on cells thus treated is irreversible.

REFERENCES

1. Alfert, M., and Geschwind, I. I. A Selective Staining Method for the Basic Proteins of Cell Nuclei. Proc. Natl. Acad. Sci. U. S., *39:* 991–999, 1953.
2. Hotta, Y., Ito, M., and Stern, H. Synthesis of DNA during Meiosis. Proc. Natl. Acad Sci. U. S., *56:* 1184–1191, 1966.
3. Ito, M., and Stern, H. Studies of Meiosis *in vitro*. I. *In vitro* Culture of Meiotic Cells. Develop. Biol., *16:* 36–53, 1967.
4. Kajiwara, K., Kim, U. H., and Mueller, G. C. Phleomycin, an Inhibitor of Replication of HeLa Cells. Cancer Res., *26:* 233–236, 1966.
5. Kihlman, B. A., Odmark, G., and Hartley, B. Studies on the Effects of Phleomycin on Chromosome Structure and Nucleic Acid Synthesis in *Vicia faba*. Mutation Res., *4:* 783–790, 1967.
6. Stern, H., and Hotta, Y. Chromosome Behavior during Development of Meiotic Tissue, *In:* L. Goldstein (ed.), The Control of Nuclear Activity, pp. 47–76. Englewood Cliffs, N. J.: Prentice Hall, 1967.
7. Tanaka, N., Yamaguchi, H., and Umezawa, H. Mechanism of Action of Phleomycin. I. Selective Inhibition of the DNA Synthesis in *E. coli* and in HeLa Cells. J. Antibiotics (Tokyo), *16:* 86–91, 1963.

Fig. 1. Cytologic changes in cells exposed to phleomycin during early meiotic stages. 1a, Leptotene-zygotene cells after 2 days of exposure to phleomycin 2 μg/ml. Type I cells are those without any condensations in the nuclei. Type II are those with small condensation regions. Type III are those with the bigger condensations. × 400. 1b, Control cells after 2 days of growth. This stage is now zygotene. The cells in the control culture were taken from the same population of cells shown under a. × 400. 1c, Leptotene-zygotene cells treated as in a. These cells show what is presumed to be a transition from Type II to Type III. The individual condensation regions are still evident. For measurements of frequency, these were considered Type II. × 1000. 1d, A control cell for comparison with cells shown in c. Compare the threadlike structure of the normal cell with that of the cells exposed to phleomycin. × 1000. 1e, Leptotene-zygotene cells exposed for 2 days to phleomycin 2 μg/ml and then transferred to inhibitor-free medium for an additional period of 3 days. The extreme condensation of the nuclei is evident. Such cells are classified as Type III. × 400. 1f, A higher magnification of a cell taken from the culture shown in e. The enlargement is intended to show the heterogeneous appearance of the condensed nucleus. Control cells cultured for the same interval of time were in the pachytene stage. × 1000. Figs. 1a–f were stained with propionic acid orcein.

Fig. 2. Cytologic changes in cells exposed to phleomycin during the later prophase stages of meiosis. 2a, Control cells explanted at zygotene and cultured for 5 days. These cells are now in late pachytene. × 400. 2b, Control cells explanted at late pachytene and cultured for 5 days. Such cells complete meiosis and form the characteristic tetrads shown here. × 400. 2c, Mid to late pachytene cells cultured for 2 days in the presence of phleomycin, 2 μg/ml, and then transferred to inhibitor-free medium. The cells shown have been cultured for an additional 6 days following transfer. The formation of cell walls as well as the condensation of nuclei is apparent. The controls for these cells reached the tetrad stage as shown under b. × 400. 2d, Higher magnification of a cell taken from the culture illustrated in c. × 1000. 2e, Cells explanted at late pachytene into medium containing 2 μg/ml of phleomycin. After 2 days of culture in this medium, the cells were transferred and grown for an additional 3 days in an inhibitor-free medium. The condensed micronuclei are clearly evident and so are the cell walls. Presumably the chromosomes separated at late pachytene and then condensed in the manner shown. No metaphases or anaphases were evident in samples examined during the interval of culture. × 400. 2f, Higher magnification of cells shown under e. Control cultures were in the tetrad stage after the same interval of incubation. × 1000. Figs. 2a–f were stained with propionic acid orcein.

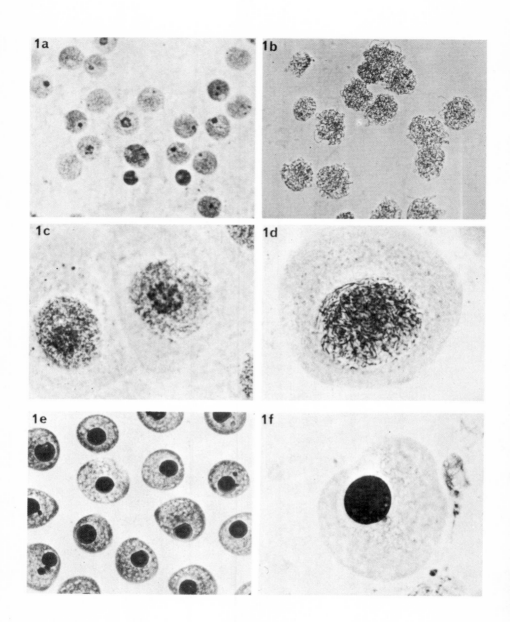

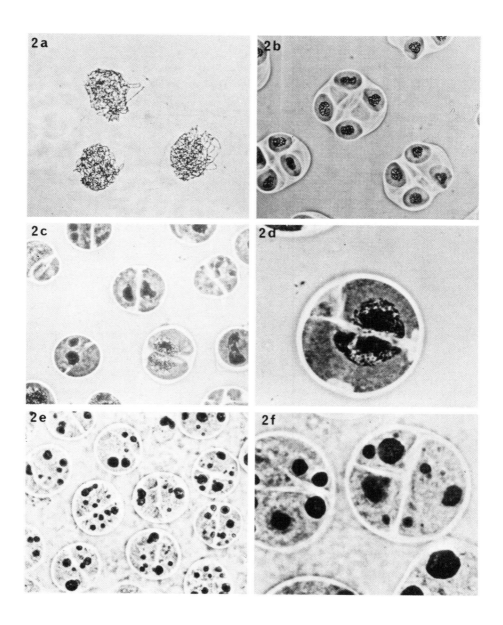

CHROMATIN-ORGANISATION DURING AND AFTER SYNAPSIS IN CULTURED MICROSPOROCYTES OF *LILIUM* IN PRESENCE OF MITOMYCIN C AND CYCLOHEXIMIDE

S. K. SEN

The significance of the universal occurrence of synaptinemal complexes (SC) at meiotic prophase has not yet been clearly established. It is understandable that their possible involvement in the process of synapsis and genetic exchange has been speculated on. It has been further realised that the G2 phase does not exist in the meiotic cycle and that little DNA synthesis takes place at meiotic prophase. Inhibition of meiotic DNA synthesis to pachynema has yielded cytological disturbances [5], including prevention of formation of SC [10]. The functional significance of DNA and protein syntheses during zygonema–pachynema has been pointed out [1, 3, 12].

Morphological information through studies on meiotic prophase chromosomes lead to visualise that at least two major organisational adjustments take place in chromatin to fit into the process of synapsis resulting in genetic exchange and to undergo meiotic division after separation of the homologues. One obvious implication of DNA and protein syntheses at meiotic prophase is the organisation of chromatin during these stages. This communication presents evidence of the organisational pattern of genetic material obtained through partial blocking of DNA synthesis by mitomycin C (M) and nuclear protein synthesis by cycloheximide (C).

Materials and Method

Leptonema and zygonema microsporocytes of lily, variety Enchantment Orange were cultured according to the methods described by Ito & Stern [4] at 20°C. Strings of meiocytes from eight buds of each length were cultured separately on four sets of experiments in presence of 2×10^{-5}M of M and 0.25 µg/ml, 0.5 µg/ml, 1 µg/ml and 2 µg/ml of C. The above-mentioned level of M is known to inhibit 76 % of DNA synthesis during zygonema–pachynema [5] and 0.2–1.0 µg/ml of C has been found to inhibit protein synthesis incompletely during zygonema–pachynema [3]. In this text, reference to 2×10^{-5}M of M and 0.25 µg/ml of C is made, as at higher doses of C meiosis often stopped altogether. Control line of the experiment was maintained from the strings of the meiocytes, one from each anther of the same bud used in other treatment lines; everyday harvesting of the materials was done for 4–5 days of culture. The degree of synchrony and the stage of meiosis were checked by squashing two or three strings from a bud. Since all the strings of the anthers of a bud were not in a strict sense synchronised, selective use of meiocytes based on light microscopic observations was done. The leptonema and zygonema had further been subdivided into substages depending upon the time of explantation and also through the degree of axial core and SC formations. Although these subdivisions were not very rigidly defined, nonetheless they helped in handling the material successfully.

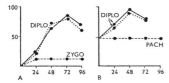

Fig. 1. Effect of presence of mitomycin C and cycloheximide during leptonema (*A*) and zygonema (*B*). *Abscissa:* Hours of culture; *ordinate:* frequency of SC. ——, Frequency (%) od SC in chromatin masses in cultured cells; – – –, mitomycin C treated cells; – – – –, cycloheximide-treated cells.

For electron microscopy the material was fixed in glutaraldehyde followed by OsO$_4$ as described by Ledbetter & Porter [6] and was embedded in Vestopal W. Sections were cut with Porter Blum mictotome and after staining with uranyl acetate and lead citrate [8] they were examined under Siemens Elmiskop I at 80 kV. The frequency of SC was determined on the basis of presence of SC in chromatin masses against their absence [9] from atleast 50 randomly selected nuclei.

RESULTS

Effects of Mitomycin C

Microsporocytes cultured in presence of *M* at leptonema after four days of development have not shown any trace of SC. Corresponding cultures of microsporocytes without *M* (control) have shown a day-to-day increase of SC till third day (pachynema). Formation of SC has been prevented when the explantation is done at mid-leptonema in presence of *M*. It is believed that the inhibition of DNA synthesis hinders continuous formation of SC throughout the length of chromosome. In fig. 1 it is indicated that the gradual increase of SC per chromatin mass is not maintained in presence of *M* as in control. Microsporocytes cultured with *M* at zygonema have also shown cessation in further formation of SC in four days of their culture. The rate of development of microsporocytes in presence of *M* is substantially affected. Slowing of the rate of meiosis has been more pronounced in the case of microsporocytes cultured at leptonema than they have been at zygonema.

Meiocytes cultured at late leptonema and or early zygonema have been found to be at pachynema on the fourth day. At this stage chromatin revealed certain interesting organisational pattern along the longitudinal axis of the ill-formed SC. Heterogeneity in chromatin condensation is marked with areas of loose strands of 250 Å in diameter. They are seen to organise in lampbrush-loop like (figs 2, 3) arrangements. The continuity of the 250 Å strands has been effectively followed through continuous sections. Transverse filaments of SC have been seen to reach to the central element in some areas where as in other areas they are loosely associated with lateral element (figs 2, 3). It cannot be established that the disclosure of clear loops of strands (250 Å) of chromatin has been possible due to inhibition in building up of axial cores in these regions or because of disturbance in normal close packing between chromatin and lateral elements through transverse filaments.

Effect of Cycloheximide

Microsporocytes cultured at leptonema and zygonema in presence of 0.25 µg/ml of cycloheximide have shown almost a similar day-to-day formation of SC as in control (fig. 1). Rate of meiosis has also been remarkably similar to the cultures maintained without *C*. Preparations of chromatin at pachynema have not revealed any disturbance in normal formation of elements of SC. Certain preparations of the microsporocytes cultured with *C* at zygonema and when harvested at diplonema have revealed (fig. 4) through serial sections, a different pattern of chromatin organisation. Chromatin strands of 250 Å have disclosed a back-and-forth sort of arrangement (fig. 4). This type of organisation has not been observed at pachynema. It has been believed that after traces of SC are dissociated gradually, chromatids arrange themselves by organising chromatin in a bach-and-forth pattern of arrangement.

DISCUSSION

Synthetic events of synapsis are beginning to appear increasingly complex although interesting, with evidence of cytologically differentiable functional significance of DNA and protein

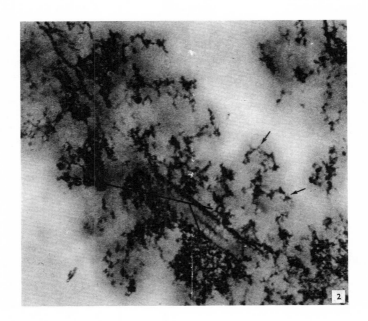

Figs 2 and 3. Mitomycin-treated cells show loops of 250 Å strands (*arrow*) and also reveal certain details of synaptinemal complexes (*SC*) with particular reference to the association of lateral elements and chromatin (*double-lined arrow*). ×42,000 and 40,000.

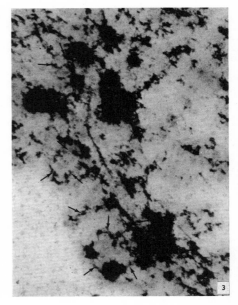

syntheses. The understanding of the involvement of SC into the synaptic chromosomes has opened up a possible line of approach to visualise the organisational pattern of chromatin taking part in synapsis for genetic exchange. Two reports are of importance in the context of this text. One of them [9] indicates as is shown here also that DNA synthesis at late leptonema and early zygonema initiates the formation of SC and the other [10] implies that distinctive protein synthesis at late zygonema to early pachynema resulting in chiasma formation [3] without affecting SC characteristics at pachynema [10]. It has been pointed out also that the effect of C at zygonema does not disrupt the organisation of SC. The significance of these syntheses in relation to synapsis, disjunction chiasma forma-

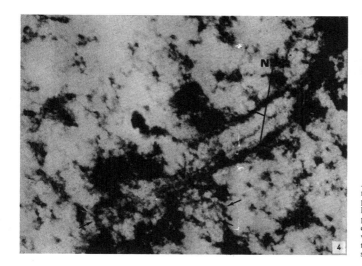

Fig. 4. Section through chromatin at early diplonema in presence of cycloheximide has revealed back-and-forth pattern of organisation of chromatin strands (*arrows*) when sometimes nonfunctional SC (*NSC*) can also be seen. ×55,000.

tion and anaphase II segregations has been discussed by previous workers [3, 5, 9, 10] and also by the present author [12].

In the light of past evidence it is believed that leptonema chromosomes organise to have axial cores which at zygonema start to involve themselves into the formation of SC [7]. It has been shown in an another communication [11] that at leptonema the axial cores are formed in each chromatid irrespective of their presence, or absence of homology. The exact stage of inhibition of DNA synthesis here leading to failure in formation of SC, is not known. Whether it prevents organisationing of the axial core or interdigitation of the transverse filaments, is still a matter of conjecture. Either or both of them may cause prevention of SC formation resulting in nonpairing. In presence of M, heterogeneity in condensation at pachynema is believed to be caused due to prevention in formation of SC. Disclosure of lampbrush-loop like arrangements of continuous chromatin strands cytologically confirms the schemes suggested earlier [7, 11, 13]. This pattern of organisation separates the two forms of chromatin of which only one participates in genetic exchange. Through inhibition of DNA synthesis M prevents formation of SC. On the other hand C does not produce the same effect, although it is known to inhibit DNA synthesis [3] considerably.

It is obvious through cytological views of chromosomes at meiotic prophase, that chromatin undergoes a gradual change in its organisation after separation of the homologues. A lampbrush-like organisation in pachynema cannot possibly be retained, as is evident from the loss of all elements of SC after termination of synapsis. It is proposed here that a gradual developmental reorganisation of the chromatin into the individual chromatids takes place during and after 'diffuse stage' [7]. The life of lampbrush state after synapsis is perhaps dependent upon the length of 'diffuse stage'. The reorganisation of chromatin after synapsis can be quite slow and gradual in some systems. The pattern of reorganisation of the chromatin into the individual chromatids attained gradually till metaphase I, is probably comparable to the pattern present in the chromatids at mitosis.

Meiocytes cultured in presence of C at zygonema have been revealed through serial sections of presence of back-and-forth folding of single strands of 250 Å fibres, at diplotene. It has been difficult to document through micrographs the continuity of the pattern of organisation of back-and-forth folding in some limited areas more than is done in fig. 4. Model of folded fibre of chromosomes has already been earlier suggested [2]. The principle of folding back-and-forth may basically constitute the pattern of organisation of chromatin threads after synapsis.

Author wishes to thank Professor J. Straub and Dr I. Anton Lamprecht for their interest and Mrs S. Sen, Miss B. Meyer and Mr D. Bock for help.

Financial grants from Alexander von Humboldt Stiftung and Max-Planck-Gesellschaft are acknowledged.

REFERENCES

1. Bogdanov, Yu F, Liapunova, N A, Sherudilo, A I & Antropova, E N, Exptl cell res 52 (1968) 59.
2. DuPraw, E J, Nature 206 (1965) 338.
3. Hotta, Y, Parchman, L G & Stern, H, Proc natl acad sci US 60 (1968) 575.
4. Ito, M & Stern, H, Devel biol 16 (1967) 34.
5. Ito, M, Hotta, Y & Stern, H, Devel biol 16 (1967) 54.
6. Ledbetter, M C & Porter, K R, J cell biol 19 (1963) 239.
7. Moens, P B, Chromosoma 23 (1968) 418.
8. Reynolds, E S, J cell biol 17 (1963) 208.
9. Roth, T F & Ito, M, J cell biol 35 (1967) 247.
10. Roth, T F & Parchman, L G, 25th Meeting of EMSA (1967).
11. Sen, S K, Bipartite axial cores in plant meiotic chromosomes. In press (1969).
12. — Nature. In press (1969).
13. Whitehouse, H L K, J cell sci 2 (1967) 9.

Peter Beaconsfield
Jean Ginsburg.

ORAL CONTRACEPTIVES AND CELL REPLICATION

SIR,—Dr. Shearman (Feb. 17, p. 325) reports the necessity of clomiphene therapy for treatment of the secondary amenorrhœa developing after oral contraceptives. These drugs have many other side-effects. We have noted that postoperative recovery is slower, and complications (including transient hypotension) are more common, in patients on long-term oral contraceptives. To translate our clinical impressions into scientific terms within a reasonable time, we devised animal experiments simulating as far as possible the situation in women taking these drugs. We investigated the effects of long-term progestogen administration on cell replication, the rate of formation of granulation tissue, and on certain metabolic processes in the vital organs with particular regard to the ramic points where metabolic-flow direction is determined. The results of these experiments are being reported with our other collaborators in appropriate journals.

We found glucose availability in the cell to be decreased in most tissues, with resultant changes in several metabolic processes—notably the rate of nucleic-acid production, with consequent effects on protein biosynthesis. The changes in protein biosynthesis prompted us to investigate the effects of oral contraceptives on the maturation of the oocyte; the results reported below relate to the experiments using norethynodrel. We used progestogens only, since the present trend of oral contraception is to use them alone, œstrogens having been incriminated as the cause of most of the side-effects; further, as pointed out by Wynn and Doar,[1] modifications in synthetic steroids may produce physiological and metabolic effects not exerted by the naturally occurring hormone.

Two groups of 10 female hooded Lister rats (œstrus cycle 4–5 days), under 4 months old and of proven fertility, received daily doses of 0·25 mg. of norethynodrel in a 2 ml. aqueous suspension for 120 days by gavage, and 20 equivalent controls were given water only by this route. After discontinuing the drug for 15 days, the ovaries from 10 treated and 10 control animals were removed under ether anæsthesia. Each ovary yielded about 20 ova, which were incubated and prepared for examination essentially by the method of Edwards.[2]

After 16 hours' incubation 90% of the ova of the control group had reached the haploid stage (figure a); only 10% failed to enter meiosis. In the treated animals, only half the ova completed normal meiosis. Some 40% did not enter meiosis: either the nuclear material was clumped together in an amorphous mass without any apparent attempt to organise into

1. Wynn, V., Doar, J. W. H. *Lancet*, 1966, ii, 715.
2. Edwards, R. G. *Nature, Lond.* 1962, **196,** 446.

chromosomes, or some unsuccessful effort had resulted in chromosomes which remained congregated at the central plate (figures b and c). In the rest, meiosis had not advanced beyond the diploid stage (figure d). This " diploid appearance " was still present in a further batch of ova incubated for 20-24 hours, suggesting that development had become arrested at this stage.

In fertility experiments, the control animals readily conceived and had 8-10 offspring per litter. After discontinuing the treatment, none of the experimental animals conceived within 30 days; only 7 conceived between 30 and 120 days, and they had 4-6 offspring per litter.

Abortuses with chromosomal anomalies as observed by Carr,[3] and reports of amenorrhœa and secondary sterility,[1,5] in women who have been on oral contraceptives for a long time seem to be very much in line with our findings in the small number of rats we have tested.

Family Planning and population control are social and economic necessities today, and oral contraception is certainly the most æsthetic method currently available. Side-effects are bound to occur with any drug, but, up to the present, standards of drug safety have equated undesirable effects strictly with the therapeutic value of the drug, taking into account the dosage and length of time for which it is given. We are now faced with the prospect of millions of healthy women taking a drug during most of their adult lives for non-essential purposes. In view of

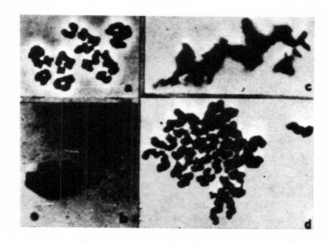

Ova of rats after 16 hours' incubation.

(a) Ovum of control rat showing haploid stage.

Ova of rats treated with norethynodrel for 120 days showing: (b) clumping of nuclear material; (c) chromosomes congregated at central plate; (d) " diploid appearance ".

3. Carr, D. H. *Lancet*, 1967, ii, 830.
4. Whitelaw, M. J., Nola, V. F., Kalman, C. F. *J. Am. med. Ass.* 1966, 195, 780.
5. Dodek, O. I., Kotz, H. L. *Am. J. Obstet. Gynec.* 1967, 98, 1065.

current limited knowledge about the fundamental effects on metabolism of oral contraceptives, their cumulative effects, and the reversibility of such changes after protracted administration, an empirical regimen of nine months on and three months off is suggested for the present.

How can we find out what influence long-term oral contraception will have, and its possible biogenetic consequences, without waiting until the end of the reproductive life of the present generation? The effects of various compounds on intracellular behaviour are studied by molecular, cell, and micro biologists, using techniques standard in their specialties. We suggest that the techniques may be adapted to the testing of drugs for studying their effects at the cellular level. One of us (P. B.) is at present preparing a schedule of drug-testing along these lines. Using such methods, results can be obtained from animals in a few months from which the likely effects in man of long-term administration of any product can be predicted with a great deal of accuracy.

Effect of 5-bromodeoxyuridine on the male meiosis in Chinese hamsters (Cricetus griseus)

ANIL B. MUKHERJEE

There are many published reports on the effect of purine and pyrimidine analogues on mitotic cells *in vitro*[2,3,5,7,8,12]. However, no published evidence could be found to demonstrate the effects of these substances on the meiotic cells *in vivo*. Since DNA synthesis has been evidenced during meiosis in plants[4] and in mammals[10] and since it is well known that 5-bromouracil and 5-BUDR are incorporated with concomitant thymine replacement, into the DNA of bacteria[3,11,12], bacteriophages[8] and mammalian cells[2,5], the following experiment, the detailed results of which will be published elsewhere, has been conducted to observe the effect of 5-BUDR on different stages of male meiosis in Chinese hamsters (*Cricetus griseus*).

5-BUDR (California Biochemical Foundation) was dissolved in balanced salt solution and 0.2 ml of the solution in the final concentration of 50 μg/ml was injected directly into the testes of 12 Chinese hamsters. At the same time, 6 animals were injected with 0.2 ml of balanced salt solution and they were regarded as controls for this experiment. At the time of the experiment the animals were 2.5 months old and were undergoing active spermatogenesis. One control and two experimental animals were sacrificed each time after 3, 6, 12, 24, 48 and 60 h of injection of 5-BUDR. Testicular preparations were made on glass slides and stained with 2% aceto–orcein stain. Spermatogonial metaphases and different phases of meiotic prophase were examined and scored for abnormality. The most frequent abnormalities were fragmentation, stickiness and extremely beaded configuration of the chromosomes. Both the control and experimental slides were coded and scored blindly by one person. The percentage of abnormal meiotic stages were calculated from at least 1000 cells counted for each treatment. Table I summarizes the data from this experiment.

HSU AND SOMERS[5] have demonstrated that 5-BUDR incorporates into the mitotic chromosomes of Chinese hamster cells *in vitro*. They have shown some specific breakage regions in the chromosome No. 1 in this organism. However, in the present investigation with spermatogonial metaphases no such specific breaks or constriction regions have been observed in that particular chromosome. There were, however, many more breaks and constrictions in chromosome No. 1 than any other chromosomes in the complement. Up to 12 h of treatment there is no significant increase in the abnormal diplotene figures. However, in 24, 48 and 60 h of treatment the abnormal diplotene figures were significantly higher than controls. Within 3 h of treatment with 5-BUDR abnormal pachytene figures were seen. Since cells which were damaged by 5-BUDR during the premeiotic "S"-period cannot migrate to pachytene stage within 3 h[1,9] and since the data presented in Table I shows more than three-fold increase in abnormal pachytene figures compared to controls it is concluded that 5-BUDR can be readily incorporated during the pachytene stages of male meiosis in Chinese hamsters. Similarly, the rise in the number of abnormal spermatogonial

---

Abbreviation: 5-BUDR, 5-bromodeoxyuridine.

TABLE I

PERCENTAGES OF ABNORMAL SPERMATOGONIAL AND MEIOTIC CELLS AFTER 5-BUDR TREATMENT

Hours after treatment	% Abnormal spermatogonial metaphases		% Abnormal pachytene nuclei		% Abnormal diplotene stages	
	a	b	a	b	a	b
3	3.4	2.8	4.1	1.2	0.0	0.2
6	12.3	3.0	8.6	1.6	0.1	0.0
12	13.6	2.6	9.2	2.0	0.1	0.0
24	18.2	1.9	8.9	0.9	1.2	0.1
48	20.1	3.1	9.3	1.8	3.2	0.3
60	20.5	2.7	9.6	2.1	4.5	0.1

a = Experimental; b = Control.

metaphases was probably due to incorporation of 5-BUDR during the premeiotic DNA synthesis period. The diplotene figures were unaffected up to 12 h of injection and after this period a rise in abnormal diplotene figures was seen. This is expected, since during diplotene there is no evidence of DNA synthesis and 5-BUDR cannot be incorporated at this stage. However, at 24 h, when the cells which were affected in pachytene migrate to diplotene they show up as abnormal diplotenes. This has been shown very nicely by Ito et al.[6] in plants by blocking DNA synthesis by specific inhibitors during the Zygonema-pachynema stages of meiosis. Presently, studies are being conducted with [³H]5-BUDR and autoradiography to see if there is any difference in the incorporation of this substance *in vivo* and *in vitro* systems.

1 CRONE, M., E. LEVY AND H. PETERS, The duration of the premeiotic DNA synthesis in mouse oocytes, Exptl. Cell. Res., 39 (1965) 678–688.
2 DJORDJEVIC, B., AND W. SZYBALSKI, Genetics of human cell lines, J. Exptl. Med., 112 (1960) 509–529.
3 DUNN, D. B., AND J. D. SMITH, Occurrence of a new base in the deoxyribonucleic acid of a strain of Bacterium coli, Nature, 175 (1955) 336–337.
4 HOTTA, Y., M. ITO AND H. STERN, Synthesis of DNA during meiosis, Proc. Natl. Acad. Sci. (U.S.), 56 (1966) 1184–1191.
5 HSU, T. C., AND C. E. SOMERS, Effect of 5-bromodeoxyuridine on mammalian chromosomes, Proc. Natl. Acad. Sci. (U.S.), 47 (1961) 396–403.
6 ITO, M., Y. HOTTA AND H. STERN, Studies of meiosis *in vitro*, II. Effect of inhibiting DNA synthesis during meiotic prophase on chromosome structure and behaviour, Developm. Biol., 6 (1967) 54–77.
7 KIT, S., C. BECK, O. L. GRAHAM AND A. GROSS, Effect of 5-bromodeoxyuridine on deoxyribonucleic acid thymine synthesis and cell metabolism of lymphatic tissues and tumors, Cancer Res., 18 (1958) 598–601.
8 LITMAN, R. M., AND A. B. PARDEE, Production of bacteriophage mutants by disturbance of deoxyribonucleic acid metabolism, Nature, 178 (1956) 529–531.
9 MUKHERJEE, A. B., Spermatogonial cell cycle in mammals, in preparation.
10 MUKHERJEE, A. B., AND M. M. COHEN, Deoxyribonucleic acid synthesis during meiotic prophase in male mice, Nature, in the press.
11 SHAPIRO, H. S., AND E. CHARGAFF, Severe distortion by 5-bromouracil of the sequence characteristics of a bacterial deoxyribonucleic acid, Nature, 188 (1960) 62–63.
12 ZAMENHOF, S., R. DE GIOVANNI AND S. GREER, Induced gene unstabilization, Nature, 181 (1958) 827–829.

MEIOSIS IN THE GRASSHOPPER: CHIASMA FREQUENCY AFTER ELEVATED TEMPERATURE AND X-RAYS

KATHLEEN CHURCH AND DONALD E. WIMBER

Introduction

Little is known about the mechanism of genetic crossing-over in higher organisms. In fact, there is even disagreement as to the time during meiosis when crossing-over occurs (Grell and Chandley, 1965; Henderson, 1966). Since chiasma formation and genetic recombination are usually considered to be manifestations of the same event (Brown and Zohary, 1955), an understanding of chiasma formation should elucidate the problem of crossing-over.

Production of chiasmata is affected by both γ- and X-radiation (Lawrence, 1961a, 1961b; Mather, 1934; Westerman, 1967). At least three radiosensitive periods affecting chiasma formation occur during meiosis and one during pre-meiotic mitoses. Lawrence (1961) demonstrated a decrease in chiasma frequency when γ-radiation was applied during the pre-meiotic DNA synthetic period and an increase when the treatment was given during late zytotene or early pachytene in both *Lilium longiflorum* and *Tradescantia paludosa*. Similar results were obtained by Westerman (1967) using X-radiation on the desert locust, *Schistocerca gregaria*, with the exception that the increase occurred when the treatment was given during leptotene-zygotene. Mather (1934) and Westerman (1967) also noted a significant increase in chiasma frequency when X-radiation was applied during pre-meiotic mitoses.

Heat shock also affects chiasma production in *S. gregaria* (Henderson, 1962, 1963, 1966) and in a grasshopper, *Goniaea australiasiae* (Peacock, 1968). A drastic decrease in chiasma frequency occurs when either *S. gregaria* or *G. australiasiae* is subjected to a temperature of 40°C. The heat-sensitive period appears to be late zygotene-early pachytene.

The present paper describes experiments designed to determine the effect of a combination of high temperature and X-radiation on chiasma production in the grasshopper *Melanoplus femur-rubrum*.

Methods and Materials

Adult male grasshoppers were collected in the field from areas surrounding Eugene, Oregon, in July and August, 1967. They were brought into the laboratory and maintained in metal cages at room temperature until the experiments were performed.

Two duplicate experiments were carried out to determine the combined effect of high temperature and X-radiation on chiasma frequency. The grasshoppers were divided into two groups in both experiments. All individuals received an abdominal injection of 5 μc H^3-thymidine in 0.005 ml water (sp. act. 6.7 C/mM). One group was then immediately subjected to 100 R of X-rays (6mA, 250 kV, 2 mm Al filtration, exposure time 13 min 20 sec) and the second group served as a non-irradiated control. Immediately after X-irradiation, both groups were placed in an insulated plexiglass cage divided into two halves by a screen to allow free passage of air to both sides of the cage. The temperature was maintained at 42°C. "Kitty litter" was scattered over the floor to reduce humidity. The hoppers were fed daily on dandelion and lettuce leaves. Samples were collected daily beginning 1 day after the initiation of the experiment. Table I indicates the number of individuals sampled each day. Mortality was rather high at such extreme temperatures and it was difficult to carry the experiments further than 9 days.

Testes were fixed in 3:1 (100% ethanol to glacial acetic acid) for 24 hours and stored in 70% ethanol in the refrigerator. One testis was stained with aceto-orcein, and squash preparations were made for determination of chiasma frequency. The other was stained by the Feulgen reaction and prepared for autoradiography.

The male of this species contains 23 chromosomes and forms 11 bivalents at meiosis (Herne and Huskins, 1935). The single X chromosome remains

TABLE I
Number of individuals sampled daily

Day	Experiment 1		Experiment 2	
	Number individuals		Number individuals	
	Heat	Heat + X-ray	Heat	Heat + X-ray
1	2	2	2	2
2	2	2	2	2
3	2	2	3	2
4	3	2	2	3
5	3	3	3	4
6	3	2	2	2
7	2	3	2	2
8	2	2	2	3
9	4	4	2	2

unpaired. Two of the bivalents can usually be distinguished by their larger size as can the two smallest bivalents. The remaining seven show a gradation in size and cannot be readily distinguished from each other.

Some terminalization of chiasmata occurs in this species. Cells at diakinesis and metaphase I contain approximately one fewer chiasmata than mid-diplotene cells. To minimize the effect terminalization would have on chiasma-frequency determination, measurements were made of the X chromosome. The X chromosome in cells entering diplotene is usually U- or S-shaped. Shortly thereafter it assumes a rod shape which becomes shorter as cells procede to metaphase I. Only those cells containing a rod-shaped X chromosome 5.5 to 6.5 μ long were used for chiasma counts. No fewer than 15 cells per male were scored and in most cases scores of up to 30 cells per individual were obtained.

A control mean chiasma frequency range was obtained by scoring 25 individuals collected from the field and maintained at room temperature in the laboratory.

Results

Autoradiography

The grasshoppers were injected with H^3-thymidine in order to label a block of cells in the pre-meiotic DNA-synthetic period at the time of injection. Meiotic development was determined autoradiographically by daily sampling. The length of the meiotic cycle and its component stages can be so determined. The results are summarized in Table II. It takes cells about 9 days to develop from the end of pre-meiotic DNA synthesis to the spermatid stage under the experimental conditions. Pre-pachytene requires 6 or 7 days for completion and pachytene, diplotene, and the two meiotic divisions take another 2 days. Table II shows that variation in the time taken to complete meiosis occurs between males sampled concurrently. For example, of four individuals sampled on day 9, three contained labeled spermatids whereas the forth had label in cells undergoing meiotic division. In all four individuals pre-pachytene

TABLE II
Timing of meiosis

Time (days after H^3-TdR injection)	Latest stage labelled			
	Pre-pachytene	Pachytene, diplotene, diakinesis	Metaphase I + II	Spermatid
2	× (3)*			
3	× (3)			
4	× (4)			
5	× (3)			
6	× (2)			
7	× (2)			
8		× (1)	× (2)	
9			× (1)	× (3)

*Number of individual grasshoppers observed.

stages were also labeled. These latter cells are most likely derivatives of cells that were synthesizing DNA one or more cell-cycles before the onset of meiosis when the animals were given H³thymidine. It should be pointed out that it was difficult to determine whether a labeled cell was in leptotene, zygotene, or early pachytene. Thus, no attempt was made to determine the length of these stages.

Chiasma Frequencies

Mean chiasma frequency per cell of both groups of males declined approximately 4 days after initial exposure to high temperatures. Figures 1 and 2 show the results of two experiments. It was discovered during the course of the first experiment (Fig. 1) that the temperature inside the experimental cage differed from bottom to top. Greater care was taken in the second experiment (Fig. 2) to maintain the temperature throughout the cage at 42°C. Despite these adjustments, the experiments gave comparable results. The mean chiasma frequencies of samples collected on days 4 to 9 from the non-irradiated group fell almost entirely outside the range of the controls. Note, however, that it was on day 6 that the chiasma frequency increased in the X-rayed lot. The mean frequency rose to within the range of the room temperature control individuals.

Our results indicate that there are at least two periods during meiosis in *M. femur-rubrum* when an environmental alteration will result in a change in chiasma frequency. The stages of meiosis affected by heat and by X-radiation can be determined by observing the autoradiographs of the timed samples. The initial reduction of chiasma frequency (Fig. 3) occurred by day 4 in both X-irradiated and non-irradiated individuals due to the high temperature. These frequencies were obtained from cells in diplotene; however, the latest stage labeled at that time was a pre-pachytene stage and labeled cells in diplotene-diakinesis did not appear until day 8 (Fig. 4). This indicates that the end of the heat-sensitive period occurs at least 3 to 4 days following pre-meiotic DNA synthesis. This would place it at a pre-pachytene stage, most likely late zygotene. Our experiments do not yield an estimate of the beginning of the heat-sensitive period. Only a high temperature pulse regime would do this. Yet, we suspect that the period of meiosis during which the high temperature is effective is of 1 or 2 days duration (cf Henderson and Peacock). An increase in chiasma frequency occurred at day 6 in the X-irradiated individuals (Fig. 5) and then declined again. The latest stage labeled in these individuals was again pre-pachytene. This radio-sensitive period occurs 2 days prior to the end of the heat-sensitive period and 1 or 2 days following pre-meiotic DNA synthesis. Thus, the radio-sensitive period is most likely leptotene and is of no more than 1 or 2 days duration.

Discussion

The reduction in chiasma frequency due to high temperature in these experiments differs from that observed by Henderson (1962, 1963, 1966) in *S. gregaria* and that by Peacock (1968) in *G. australiasiae*. Both Henderson and Peacock observed more drastic reductions in chiasma frequencies which reached almost total univalancy with prolonged heat treatment. No univalents

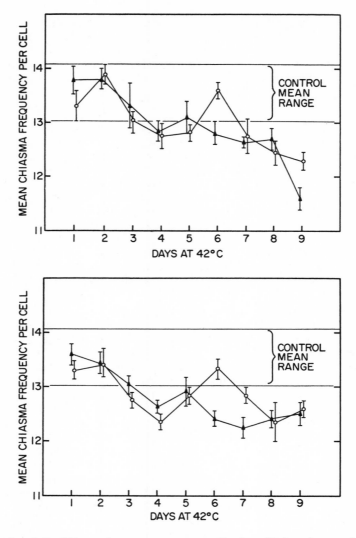

Figs. 1 and 2. Chiasma frequency response with time. Each point on the graphs (duplicate experiments) represents mean and standard errors of chiasma counts made on two to four individuals. The mean chiasma frequency of each of 25 control grasshoppers kept at room temperature was obtained. Solid parallel horizontal lines demarcate the range of these control means. ▲ = heat (42°C) treated individuals, O = heat + X-irradiated individuals.

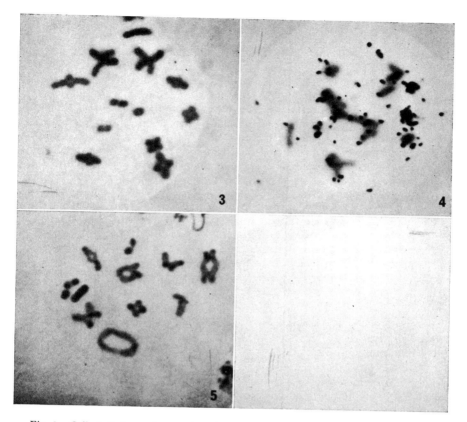

Fig. 3. Cell at diplotene in a male 4 days after subjection to heat and X-rays showing reduction in chiasma frequency. Only 11 chiasmata are present. Fig. 4. Autoradiograph of a cell at diplotene-diakinesis in an 8-day male. Fig. 5. Diplotene in a male after 6 days of treatment (heat + X-rays) showing a reversal of chiasma frequency to 14.

were observed in any of the *M. femur-rubrum* sampled. The initial chiasma frequency of *M. femur-rubrum* is much lower than that of *S. gregaria* or *G. australiasiae*. The majority of the bivalents contain only one chiasma under normal conditions and these bivalents appear to be unaffected by the heat. Three or four of the larger bivalents contain two and rarely three chiasmata. These bivalents with more than one chiasma are the ones altered by heat. Henderson (1963) attributes the heat response to complete asynapsis or desynapsis. It is obvious that heat treatment did not cause asynapsis in the present experiments since at least one chiasma per bivalent was observed in all heat-treated cells. We have also attempted to produce univalents by temperatures as high as 47°C and have failed. The effect observed by Henderson occurred at late zygotene or early pachytene. In the present experiments it was technically difficult to determine the precise heat sensitive stage, but the results indicate it was a pre-pachytene stage, most likely zygotene.

These data argue for two qualitatively different types of chiasmata: one whose presence is obligatory and which may function mainly to ensure regular co-orientation and disjunction of the chromosomes. The other type has a more flexible pattern and is more amenable to environmental influences. One might speculate that because of the obligatory nature of the first type, it plays only a minor role in enhancing genetic variability. The other type, however, being less canalized in its occurrence, may provide the genetic variation assumed in most populations. Thus, the males in our population of *M. femur-rubum* showed predominately the obligatory chiasmata with only a few of the optional type. The Henderson and Peacock males, on the other hand, displayed mainly the optional chiasmata with few or none of the others.

Both γ- and X-rays have been shown to alter chiasma frequency in plants and insects (Lawrence, 1962a, 1962b; Westerman, 1967; Mather, 1934). Westerman (1967) includes an excellent review of this literature. Our results agree with those of Westerman in that we observed an increase in chiasma frequency when X-radiation was delivered during leptotene and possibly early zygotene. The end of the X-ray-sensitive period did not coincide with the end of temperature sensitivity but came earlier in meiosis. We did not observe a second sensitive period occurring in pre-meiotic DNA synthesis as did Westerman, during which chiasma frequency was depressed; this is probably due to the design of our experiments. The chiasma frequency in the heat-treated individuals was lowered to such an extent that it would be difficult to pick up any further decrease due to X-irradiation without producing univalents; this we found impossible to do in our material.

Peacock (1968) has determined from a series of beautiful experiments in *G. australiasiae* that the heat-sensitive period (early pachytene) is precisely the period when homologues physically exchange genetic material. Lowering the chiasma frequency by heat treatment also lowers the frequency of exchange. This furnishes dramatic evidence that chiasmata and crossing-over are, indeed, casually related. With this information then, how can we explain the observation that there are two and perhaps more time periods during meiosis when the frequency of chiasma can be altered? We suggest three possibilities which may not be mutually exclusive: 1) There may be more than 1 time period of meiosis during which chiasmata are formed, 2) the high temperature and the X-rays may each be affecting different sites within a series of events leading to chiasma production, and 3) there may be more than one type of chiasma. All of these possibilities are compatible with our results and with those of Westerman (1967) which indicate the chiasmata on different bivalents, and even chiasmata within the same bivalent, may respond differently to physical changes in the environment. Along this line, Grell (1965) has formulated a hypothesis based on genetic evidence from *Drosophila* where chromosomes pair twice during meiosis. Grell suggests that chromosomes are first involved in exchange pairing and those homologues which fail to exchange chromosome material enter a distributative pool and pair a second time to insure proper segregation of non-crossover chromosomes. The two hypothesized pairing events are separated in time. If both were events preceding formation of chiasmata, this might explain the existence of two of the time periods when chiasma frequencies could be altered. It is difficult to conceive of a model where one type of chiasma is the result of a crossover event and a second or third type the result of some other unknown event or events. Nevertheless, we do know that not all chiasmata respond similarly to experimental manipulation.

Acknowledgements

This work has been supported by a Public Health Service Research Career Development Award (GM 8465), a PHS Research Grant (GM 11702) and a PHS Training Grant (GM 373) all from the National Institute of General Medical Sciences.

The authors wish to thank Miss Regina B. Ross for her excellent technical assistance.

References

Brown, S. W., and Zohary, D. 1955. The relationship of chiasmata and crossing-over in *Lilium formosanum*. Genetics **40**: 850-873.

Grell, Rhoda F. 1965. Chromosome pairing, crossing-over and segregation in *Drosophila melanogaster*. Nat. Cancer Inst. Monogr. **18**: 215-242.

Grell, R. F., and Chandley, A. C. 1965. Evidence bearing on coincidence of exchange and DNA replication in the oocyte of *Drosophila melanogaster*. Proc. Nat. Acad. Sci. U.S. **53**: 1340-1346.

Hearne, E. Marie, and Huskins, C. Leonard. 1935. Chromosome pairing in *Melanoplus femur-rubrum*. Cytologia **6**: 123-147.

Henderson, S. A. 1962. Temperature and chiasma formation in *Schistocerca gregaria*. II. Cytological effects of 40°C and the mechanism of heat-induced univalence. Chromosoma (Berl.) **13**: 437-463.

Henderson, S. A. 1963. Temperature and chiasma formation in *Schistocerca gregaria*. I. An analysis of the response at a constant 40°C. Heredity **18**: 77-94.

Henderson, S. A. 1966. Time of chiasma formation in relation to the time of DNA synthesis. Nature **211**: 1043-1047.

Lawrence, C. W. 1961a. The effect of irradiation of different stages of microporogenesis on chiasma frequency. Heredity **16**: 83-89.

Lawrence, C. W. 1961b. The effect of radiation on chiasma formation in *Tradescantia*. Radiation Botany **1**: 92-96.

Mather, K. 1934. The behavior of meiotic chromosomes after X-irradiation. Hereditas **19**: 303-322.

Peacock, W. J. 1968. Replication, recombination and chiasmata in *Goniaea australiasiae* (Orthoptera:Acrididae). (to be published).

Westerman, M. 1967. The effect of X-irradiation on male meiosis in *Schistocerca gregaria* (Forskal). Chromosoma **22**: 401-416.

AUTHOR INDEX

Beaconsfield, Peter, 199
Bobrow, M., 14, 23
Calhoun, Ford, 130
Chiarelli, Brunetto, 19
Church, Kathleen, 202

Davis, Louise M., 149

Flaek, Arthur, 19

Ginsburg, Jean, 199
Gondos, Bernard, 104

Hayes, Susan, 10
Hotta, Yasuo, 181
Howe, H. Branch, Jr., 130

Juberg, R. C., 149

Kirilova, Maria, 47

Lavappa, K. S., 159

Melnyk, John, 10
Mohamed, Aly H., 169

Moutschen-Dahmen, J., 176
Moutschen-Dahmen, M., 176
Mukherjee, Anil B., 202

Pearson, P. L., 14, 23
Pogosianz, Helen E., 78

Raicu, Petre, 47
Ramel, Claes, 141
Rucci, Alfred J., 10

Sen, S. K., 194
Solari, Alberto J., 27, 53, 70, 90
Stern, Herbert, 181
Stevens, Margaret, 166
Szollosi, Daniel, 111

Thompson, Havelock, 10

Vanasek, Frank, 10
Vladescu, Barbu, 47

Wimber, Donald E., 204

Yerganian, George, 159

Zamboni, Luciano, 104

KEY-WORD TITLE INDEX

Aging in Fallopian Tube, 111
Asci, Genetic Analysis of Eight-spored, 130

Cell Replication, Effect of Oral Contraceptives on, 199
Chiasma Frequency, 204
Chromatin-organization during Synapsis, 194
Chromosomal Axes, Behavior during Diplotene, 90
Chromosomal Changes in Maize, 169
Chromosome, Association of Y Short Arm with X, 23
Chromosome Distribution in Metaphase Plates, 47
Chromosome, Fluorescent Staining of Y, 14
Chromosome, Spatial Relationship of Y to X, 27
Chromosome, Transmission of Extra, 10
Chromosomes, Changes in Sexual, 70
Chromosomes, Human Meiotic, 19
Chromosomes, Ultrastructural Evolution of Sexual, 53
Combined Treatment Fast Neutrons-FUDR, 176
Cricetus Griseus, 202
Curly Inversions, Effect of, 141
Cycloheximide Presence, 194

Djungarian Hamster, 78
Drosophila melanogaster, 141

5-bromodeoxyuridine, Effect of, 202
47,XYY Karyotype Subjects, 10

Gene E, Production of Eight-spored Asci, 130
Germ Cell Differentiation, Synchronization of, 104
Grasshopper, 204

Human, 14, 19, 23
Hydrogen Fluoride Gas, Effect on Chromosomes, 169

Induction of Spawning and Maturation, 166
Intercellular Bridges, 104

Latent Meiotic Anomalies, 159
Lilium, 194

Maize, 169
Mesocricetus newtoni, 47
Mitomycin C Presence, 194
Morphological Changes in Mouse Eggs, 111
Mouse, 27, 53, 70, 90, 111
Mutagenic Agent, Ancestral Exposure to, 159

Neurospora tetrasperma, 130
Nondisjunction, Genetic Control of, 149

1-methyladenine Treatment, 166
Oral Contraceptives, 199

Phleomycin, Effect on Cells, 181

Rabbit, 104
Recombination, Intrachromosomal Effects on, 141

Starfish, 166

Ultrastructural Evolution of Sex Chromosomes, 53

Vicia faba, 176